SOLVING LARGE SCALE MINIMUM SPANNING TREE PROBLEMS AND TRANSPORTATION PROBLEMS USING A MODIFIED ANT COLONY ALGORITHM

SOLVING LARGE SCALE MINIMUM SPANNING TREE PROBLEMS AND TRANSPORTATION PROBLEMS USING A MODIFIED ANT COLONY ALGORITHM

K.P. Oshan Niluminda
E.M.U.S.B. Ekanayake

ELIVA PRESS

ISBN: 978-99949-8-435-0

Cover Design: Eliva Press
Cover Image: Ingimage
Email: info@elivapress.com
Website: www.elivapress.com

This Eliva Press imprint is published by the registered company Eliva Press Global Ltd. part of Eliva Press S.R.L. Publishing Group

The registered company address is: Pope Hennessy Street Level 2, Hennessy Tower Port Louis, Mauritius
Eliva Press S.R.L. Publishing Group address is: Bulevardul Moscova 21, Chisinau, Moldova, Europe

SOLVING LARGE SCALE MINIMUM SPANNING TREE PROBLEMS AND TRANSPORTATION PROBLEMS USING A MODIFIED ANT COLONY ALGORITHM

K. P. O. Niluminda – ASP/2016/2017/061

Dissertation submitted in partial fulfillment of the requirements for the award of the degree of Bachelor of Science (Hons.) in Industrial Mathematics.

Department of Physical Sciences
Rajarata University of Sri Lanka.

2023

Table of Contents

Abstract

A spanning tree is a specific type of graph G. In a connected undirected graph G, a spanning tree is created when all of the vertices are connected by the least amount of edges possible. There could be a number of these spanning trees for a graph. The "Minimum Spanning Tree" is a connected, edge-weighted, undirected network with all of its vertices linked, no cycles, and the least amount of total edge weight. Researchers have created several algorithms to find the Minimum Spanning Tree. Famous algorithms designed to determine the Minimum Spanning Tree of an undirected connected graph include Prim's and Kruskal's. The Ant Colony Optimization Algorithm is a probabilistic method for handling computing issues that may be simplified to determine the best path via graphs. This method was developed utilizing the foraging technique employed by ants to locate a route from their colony to a food source.

The Transportation Problem is concerned with identifying routes to deliver the items to numerous destinations to fulfill the needs of the many destinations and supply quantities from various sources. In the literature, several methods for resolving Transportation Problem have been established. Two-step strategies such as an Initial Basic Feasible Solution and an Optimal Solution are possible in Transportation Problem. The Modified Distribution Method or the Stepping Stone Method may be used to find an Optimal Solution for the Transportation Problem, while the North-West Corner Method, Least Cost Method, Vogel's Approximation Method, and other methods can be used to find an Initial Basic Feasible Solution.

This study proposed a novel approach to finding the Minimum Spanning Tree of undirected, connected graphs using the Modified Ant Colony Optimization Algorithm. Then apply this algorithm to find the Minimum Spanning Trees of the Indian Domestic Air Network. Finally, the proposed method developed to find the Initial Basic Feasible Solution to both classical and prohibited route Transportation Problems. In each step obtained results compared with other existing methods.

Keywords: Minimum Spanning Tree, Ant Colony Optimization, Indian Domestic Air Network, Classical Transportation Problem, Prohibited Route

Acknowledgment

My heartfelt gratitude goes to my supervisor Dr. Uthpala Ekanayake, Senior Lecturer, Department of Physical Sciences, Faculty of Applied Sciences, Rajarata University of Sri Lanka for his valuable advice, support, constant encouragement, and guidance given to complete this project successfully.

I like to take this opportunity to thank Prof. E. M. R. K. B. Edirisinghe, Dean of the Faculty of Applied Sciences at Rajarata University of Sri Lanka, Dr. (Ms.) Harshani O. Wijewardane, Head of the Department of Physical Sciences, and all other academic staff members who helped in various ways to complete this work.

Finally, I express my heartfelt gratitude to my family members and friends who gave me continuous support to complete this project successfully.

List of Tables

List of Figures

List of Abbreviations

AAI – Airports Authority of India

ACO – Ant Colony Optimization

AOCM – Average Opportunity Cost Method

APLCP – All-Pairs Least Cost Problem

ATM – Allocation Table Method

BCTP – Balanced Classical Transportation Problem

BOTP – Bi-Objective Transportation Problem

BPRTP – Balanced Prohibited Route Transportation Problem

CAM – Column Minima Method

CTP – Classical Transportation Problem

DENSEA – Duplicate Elimination Non-dominated Sorting Evolutionary Algorithm

IBFS – Initial Basic Feasible Solution

ICM – Implied Cost Method

ICVM – Inverse Coefficient of Variation Method

IDAN – Indian Domestic Air Network

IDTP – Interval Data-based Transportation Problem

iLCM – Improved Least Cost Method

IVAM – Improved Vogel's Approximation Method

IZPM – Improved Zero Point Method

LCM – Least Cost Method

MACOA – Modified Ant Colony Optimization Algorithm

MMSTP – Multi-objective Minimal Spanning Tree Problem

MODI – Modified Distribution

MOEA – Multi-Objective Evolutionary Algorithm

MST – Minimum Spanning Tree

MWST – Minimum Weight Spanning Tree

NEWM – New Method

NMST – Neutrosophic Minimal Spanning Tree

NSGAII – Non-dominated Sorting Genetic Algorithm II

NWCM – North West Conner Method

OS – Optimal Solution

PRTP – Prohibited Route Transportation Problem

SDM – Standard Deviation Method

RAM – Row Minima Method

TP – Transportation Problem

UCTP – Unbalanced Classical Transportation Problem

UMMSTP – Uncertain Multi-objective Minimum Spanning Tree Problem

UPRTP – Unbalanced Prohibited Route Transportation Problem

VAM – Vogel's Approximation Method

List of Appendices

CHAPTER 1

Introduction

In the real world, people frequently seek employment at the lowest possible wage while also taking into account their travel time and distance to work. In a travel graph, the distance traveled refers to the point of touch with the edge that on a spanning tree contains the fewest types of labels. When each edge in a graph is assigned a label from a finite label set rather than a weight, the minimum labeling spanning tree problem seeks to determine the spanning tree that has the fewest different types of labels. One of the most common and well-known problems in combinatorial optimization is the Minimum Spanning Tree (MST) problem.

An undirected connected graph's spanning tree is a sub-graph of that graph that contains all of the vertices with the fewest amount of edges. A spanning tree cannot exist if a vertex is missing. Weights may or may not be applied to the edges. Since it can't be stretched to all of its vertices, a disconnected graph lacks a spanning tree. The spanning tree will become disconnected if you take one of the edges out. By expanding the spanning tree by one edge, a loop is created. An edge that joins two vertices is known as a loop. There can be $n^{(n-2)}$ spanning trees in a fully connected undirected network, where n is the total number of vertices. The term "complete graph" refers to a graph that connects every pair of vertices and edges. There is always at least one spanning tree in a connected and undirected graph, but there are none in a disconnected graph. By deleting edges from a complete graph, a spanning tree may be created. The spanning tree with a weight that is less than or equal to the weight of all other possible spanning trees is referred to as the "Minimum Spanning Tree (MST)." There can only be one spanning tree with a weight greater than zero when every vertex in a network is connected and free of cycles. The weights of all the edges in the tree may be added together to determine the overall cost of the MST. An MST is used in several applications, including telecommunications systems, civil network design, clustering, water supply networks, and transportation systems. The minimum spanning trees of the graph may be found using a variety of approaches. Depending on how they are implemented, these algorithms can be classified into two categories. There are two types of MST: Line (Edge)-based and Node (Vertices)-based.

The Ant Colony Optimization (ACO) Algorithm is a probabilistic approach for dealing with computing problems that may be reduced to finding the optimum path via graphs. This algorithm was created using the foraging strategy used by ants to find a path between their colony and a source of food. Finding the most efficient path between the source and the destination is made easier by the approach used. One probabilistic method for resolving a variety of optimization issues is the ACO Algorithm. ACO algorithm, used in computer science and operations research, is a stochastic method for resolving computing issues that may be simplified to discover efficient routes across networks. Multi-agent techniques inspired by the behavior of actual ants are represented by artificial ants. Biological ants' pheromone-based communication is frequently the dominating paradigm. For many optimization problems requiring some kind of graph, such as vehicle routing and internet routing, combinations of artificial ants and local search algorithms have emerged as the go-to technique. One family of optimization techniques that are based on the behavior of an ant colony is called ACO. By navigating a parameter space that represents all potential solutions, artificial "ants" (such as simulation agents) find the best answers. While scouting their surroundings, real ants leave behind pheromone trails that guide one another to resources. To help more simulated "ants" find better solutions during subsequent simulation rounds, the simulated "ants" similarly record their locations and the quality of their answers. The bees algorithm, which more closely resembles the foraging habits of the honey bee, another social insect, is one variation of this strategy.

Linear programming is extensively used in the physical transportation of goods and services from numerous supply centers to multiple demand centers. A Transportation Problem (TP) stated in terms of an LP model may also be solved using the simplex method. Even if a TP contains several variables and restrictions, utilizing simplex approaches requires a lot of time. The Stepping Stone Method and the Modified Distribution (MODI) Method are two algorithms that have been used to find the optimal solution for the TPs. The TP's structure is made up of several shipping routes connecting diverse supply locations with various demand centers. The objective of choosing the shipping routes between supply and demand centers is to satisfy the demand for a specific number of goods or services at each destination location while spending the least amount of time and/or money on transportation. The entire cost of transporting a homogeneous commodity (item) from supply to demand areas can be decreased by applying transportation algorithms. But it may also be utilized to increase the overall worth or utility.

The TP on Prohibited or Unacceptable Routes (PRTP) is a specific instance of TP. When shipments from one of the sources cannot reach one or more destinations, it occurs. When tackling these types of minimization problems, we assume that a confined cell will have a very high unit cost (+M or $+\infty$) and a very low cost (-M or $-\infty$).

1.1 Basic Definitions

1.1.1. Graph

The term "graph" (G) refers to a collection of finitely many vertices or nodes (V) linked by edges or arcs (E). A graph, on the other hand, is made up of two components: vertices (V) and edges (E).

1.1.2. Degree of the Vertex

The number of edges that intersect a vertex to determine its degree in an undirected graph. If the graph has loops, the degree of the loop at each vertex is increased by two. Vertex degree $V(G)$ is indicated by deg (v) or $\delta(v)$.

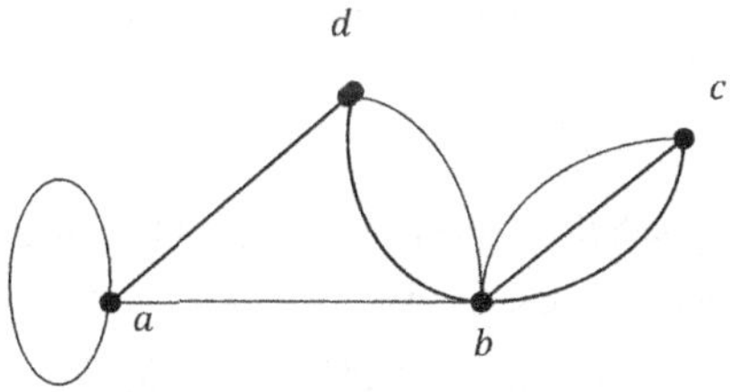

Figure 1.1: Connected Graph

In Figure 1.1:

$V(G) = \{a, b, c, d\}$

$E(G) = \{ab, aa, bc, bc, bc, bd, bd, ad\}$

$\delta(a) = 4, \delta(b) = 6, \delta(c) = 3, \delta(d) = 3$

1.1.3. Adjacent Vertices

Two vertices in graph G (V, E) are said to be adjacent if they are connected by an edge, and two edges are described as adjacent if they share at least a single vertex.

1.1.4. Subgraphs

A graph containing vertices that are part of V(G) and edges that are part of E(G) is referred to as a subgraph of a graph G (V, E). You may create subgraphs of a graph by eliminating its vertices and edges.

1.1.5. Regular Graph

If every vertex in a graph has the same degree as a simple graph, the graph is said to be regular. Meaning that $\delta(v)$ is constant for every $v \in V(G)$.

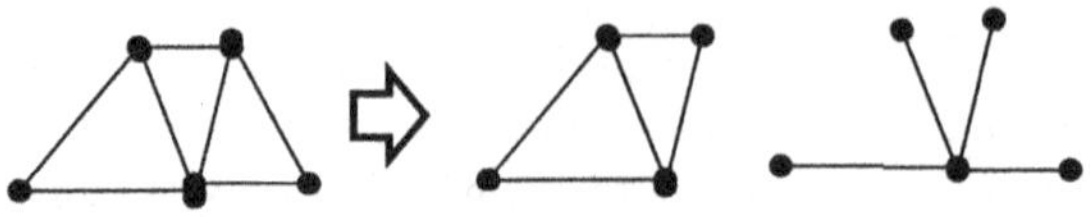

Figure 1.2: Edges & Vertices deleting

1.1.6. Loop

If an edge in a graph has the same start and end vertex, it is said to be a loop (joins a vertex to itself).

1.1.7. Cycle

The term "cycle" refers to any path that starts and finishes at the same vertex. Figure 3 below shows a graph with a loop in G_1 and a cycle in G_2.

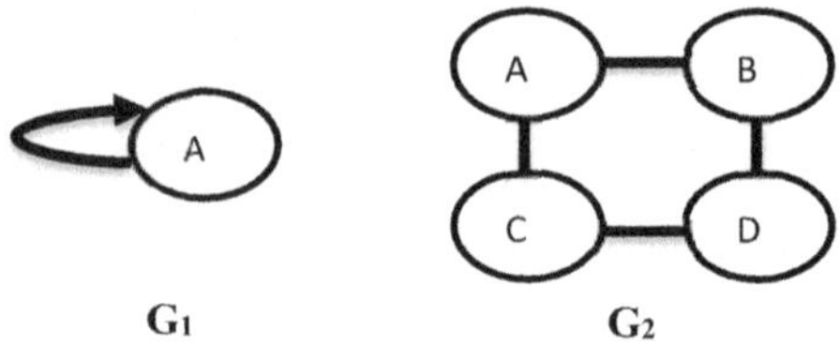

Figure 1.3: Loop & Cycle

1.1.8. Directed and Undirected Graph

Undirected graphs have edges without orientation. It is possible to go down each edge in both directions, demonstrating a two-way interaction. This graphic shows a straightforward undirected graph with three nodes and three edges. In directed graphs, each edge has a direction.

1.2. Spanning Tree

A spanning tree is a subgraph of an undirected connected graph.

1.3. Minimum Spanning Tree (MST)

The spanning tree with the least total edge weight is the minimum spanning tree. The weight of a spanning tree is the total of the weights given to its edges.

1.4. Minimum Weight Spanning Tree (MWST)

The tree that has the fewest total edge weights is known as the MST. It consists of every vertex from the same graph. The weight of a minimum spanning tree is the total of the weights given to each of its edges.

The spanning tree is technically defined as $G(V, E')$ and $E' \subseteq E$ if a connected undirected graph G (V, E) exists. Equation (1) defines the weight of the MST as follows:

$$w(T) = \sum_{e \in T} w(E') \tag{1}$$

In a weighted undirected network, the smallest weight-spanning tree may be found using either Prim's method or the Kruskal method.

1.5. Prim's Algorithm

Prim's algorithm is a popular greedy algorithm. Below is a description of how this method is implemented.

Step 1: First, pick a vertex at random.

Step 2: Pick the edge adjacent to that vertex with the lowest weight.

Step 3: Identify every edge that joins the tree to new vertices.

Step 4: Select the edge that has the least weight relative to the other edges and add it to the existing tree.

Step 5: If adding that edge results in a cycle, drop it and choose the next edge with the minimum weight.

Step 6: Continue performing steps 3-5 until the MST has been built and all vertices have been taken into account.

1.6. Kruskal's Algorithm

Kruskal's Algorithm is another renowned greedy algorithm. Kruskal's approach can only be applied to weighted, connected, and undirected graphs. The following are the stages for applying Kruskal's algorithm:

Step 1: Arrange all of the edges in ascending weight order.

Step 2: Use the smallest weighted edge to join the points on the graph.

Step 3: If adding an edge results in the creation of a cycle, remove it and go to the edge with the least amount of weight.

Step 4: Add edges continuously until the MST is created and all of the vertices are connected.

1.7. Ant Colony Optimization (ACO) Algorithm

Some kinds of ants in the natural world initially roam before returning to their colony while leaving pheromone trails after obtaining food. If additional ants discover such a road, they are more likely to follow it rather than continue traveling randomly, returning and re-establishing it if they ultimately locate food.

However, the pheromone trail begins to disappear with time, weakening the attraction it once had. The longer it takes an ant to walk down the trail and return, the longer it has for the pheromones to dissipate. In contrast, a short path is marched over more frequently, leading to a higher pheromone density on shorter paths than on longer ones. Another benefit of pheromone evaporation is that it prevents convergence to a locally optimum solution. Without any evaporation, the early ants' trails would often be too alluring for the subsequent ants to follow. The investigation of the potential solutions would thus be limited. Pheromone evaporation has a significant impact on artificial systems but is unknown in genuine ant systems. The end effect is that additional ants are more likely to follow a path discovered by one ant when it discovers a favorable (i.e., short) route from the colony to a food source. This positive feedback cycle eventually results in a large number of ants traveling along the same route. The goal of the ant colony algorithm is to imitate this behavior by simulating "simulated ants" that move around the graph that represents the issue to be solved.

A population-based search method for determining the ideal path that is based on ant behavior is known as an ant colony optimization algorithm. Ants move from their nest (let's say, a city i) to a food source using pheromones. The probabilistic transition rule and pheromone update rule are applicable if an ant k is currently located in city i and wishes to relocate to city j.

1.7.1. Transition Rule

Equation (2) shows the probabilistic transition rule for the ACO algorithm.

$$P_{ij}^{k} = \begin{cases} \frac{[\tau_{ij}]^{\alpha}[\eta_{ij}]^{\beta}}{\sum_{z \in Allowed_k}[\tau_{iz}]^{\alpha}[\eta_{iz}]^{\beta}} & j \in Allowed_k \\ 0 & ; Otherwise \end{cases} \qquad (2)$$

The quantity of pheromone on component between cities i and cities j is indicated by τ_{ij}, and its heuristic value is denoted by η_{ij}, where P_{ij}^{k} is the probability branch of kth ant from i^{th} city to j^{th} city. Both α and β are parameters $(0 \leq \alpha, \beta \leq 1)$.

1.7.2. Pheromone Update Rule

Equations (3), (4), and (5) show how the local update rule for pheromone trails is implemented for each route.

$$\tau_{ij}(t+1) = \rho . \tau_{ij}(t) + \Delta\tau_{ij} \qquad (3)$$

$$\Delta\tau_{ij} = \sum_{k=1}^{z} \Delta\tau_{ij} \qquad (4)$$

$$\Delta\tau_{ij}^{k} = \begin{cases} \frac{\omega}{d_k} & if\ ant\ k\ travels\ on\ city\ (i,j) \\ 0 & ;Otherwise \end{cases} \quad (5)$$

Here, d_k stands for the optimum path's distance and t is time. The parameter ω controls how much pheromone is deposited and is typically set to 1.

1.8. Indian Domestic Air Network

1.8.1 Domestic Airport in India

India is a significant Asian nation with a developing economy that is quickly on its way to becoming the most advanced. Proper airports are stated to play a key role in boosting the economy and growth of the nation to retain its position in the Asian subcontinent and the rest of the world.

1.8.2. Airports in India

Roads, rivers, and airways are just a few of the transportation systems that are essential to the state's development. Today, there is a critical demand for quicker transportation that can facilitate faster commuting and business. Both local and foreign airports are widely spread throughout India. They assist in meeting the needs of tourists and travelers who come to the country as well as those who travel inside it. There are two or more domestic airports and at least one significant international airport in every state. Every major city in India, including the capital cities, has at least one domestic airport. There are also approximately 12 international airports. Around 80 domestic airports link the huge nation of this superpower. All airports have upgraded facilities and are well-equipped with cutting-edge technology. Each airport has a name and unique code.

1.8.3. Uses of the Airports

In the modern world, both local and international airports are used by millions of people for both long and short trips since they are the most convenient and efficient means of transportation. Many businesspeople and tourists now prefer flying since it allows them to get where they're going fast and comfortably. Fast transportation has several advantages for both the person and the nation's economy. Except for the difficult economic times, the civil aviation sector has had phenomenal growth, and passengers are flying more often. Both government-run and privately-owned airlines offer the amenities that travelers require. Over the past few years, there have been significant changes, and services have also significantly improved. Due to market competition, all domestic flights are interconnected, provide excellent rates, and have reasonable fees. Domestic and international flights are offered by several carriers, including Jet Airways, Sahara Airlines, Air India, and Indian Airlines.

1.9. Transportation Problem (TP)

Transporting people and products from one location to another is known as transportation. The utilization of transportation is dependent on our requirement to transfer items from their available location to their used location. Humans transport things, commodities, and ideas through a variety of techniques. The four main modes of transportation are land, train, sea, and air.

The transportation problem is one of the problems the companies experience. It initially referred to the challenge of efficiently shipping or carrying goods from the point of production to the final destination while balancing supply and demand. Specifically created for models with a linear goal and constraint functions, it belongs to a particular class of linear programming techniques. They can also be used in other operational contexts, including Optimizing a network, Organizing processes, Time management and scheduling, Managing inventory, Resource management software, and Optimization of the path.

1.9.1. Sources and Destinations

Demanding nodes, also known as destination nodes $(D_1, D_2, \dots, D_n)$ in a TP, are the places where items are provided from sources, while sources $(S_1, S_2, \dots, S_m)$ are the spots in a TP where commodities are dispersed to the necessary points.

1.9.2. Supply Limit and Demand Requirement

The word "source's supply limit" refers to the amount of a product needed to satisfy demand at each demand node, while the term "demand requirement" refers to the amount of a good needed to satisfy a demanding node's requests.

1.9.3. Balanced & Unbalanced Transportation Problem

If the entire supply and demand are equal, a TP is said to be in balance.

i.e. $[Total\ Supply\ (\sum_{i=1}^{m} a_i) = Total\ Demand\ (\sum_{j=1}^{n} b_j)]$

When the overall supply and demand differ, an unbalanced TP occurs.

i.e. $[Total\ Supply\ (\sum_{i=1}^{m} a_i) \neq Total\ Demand\ (\sum_{j=1}^{n} b_j)]$

1.9.4. Model Representation of the Transportation Problem

Consider a scenario in which a specific product is typically manufactured at m production facilities, known as sources, denoted by $S_1, S_2, \dots, S_m$ with respective capacities of $a_1, a_2, \dots, a_m$ and transported to n distribution centers, known as destinations, denoted by $D_1, D_2, \dots, D_n$ with respective needs $b_1, b_2, \dots, b_m$.

Assume that x_{ij} and c_{ij}, respectively, represent the quantity transferred and the transit expense from the ith source to the jth destination $i = 1,2, \dots, m$ & $j = 1,2, \dots, n$.

Mathematical Formulation of TP in standard form:

Objective Function:

Min Z = $\sum_{i=1}^{m}\sum_{j=1}^{n} x_{ij}c_{ij}$

Subject to:

$\sum_{j=1}^{n} x_{ij} = a_i, \quad i = 1, 2, \dots, m$

$\sum_{i=1}^{m} x_{ij} = b_j, \quad j = 1, 2, \dots, n$

$x_{ij} \geq 0 ; \forall\ i = 1, 2, \dots, m \ \& \ j = 1, 2, \dots, n$

Table 1.1: Display of the transportation problem in a table

Destination → / **Source** ↓	D_1	D_2	⋯	D_n	**Supply (a_i)**
S_1	c_{11}	c_{12}	⋯	c_{1n}	a_1
S_2	c_{21}	c_{22}	⋯	c_{2n}	a_2
⋮	⋮	⋮	⋯	⋮	⋮
S_m	c_{m1}	c_{m2}	⋯	c_{mn}	a_m
Demand (b_j)	b_1	b_2	⋯	b_n	

1.9.5. Solutions to Transportation Problems

There are two solutions to TPs. Such as the Initial Basic Feasible Solution (IBFS) and Optimal Solution.

Initial Basic Feasible Solution (IBFS)

The term "Initial Basic Feasible Solution (IBFS)" refers to the overall number of allocations made by a TP to fulfill m + n – 1. Here m means the total number of sources and n means the total number of destinations.

Optimal Solution

The term "optimal solution" refers to a situation when IBFS are at their minimum level.

1.9.6. Methods to Solve Transportation Problem

There are so many methods proposed to solve TPs in literature. Here mention some of the algorithms currently use to find IBFS and Optimal Solutions of TPs.

- Initial Basic Feasible Solution (IBFS)
    - North-West Conner Method (NWCM)

  - Least Cost Method (LCM)
  - Vogel's Approximation Method (VAM)
  - Row Minima Method (RAM)
  - Column Minima Method (CAM)
- Optimal Solution
  - Modified Distribution (MODI) Method
  - Stepping Stone Method

North-West Conner Method (NWCM)

1. Select the North-west (i.e., upper left) corner cell of the table and allocate the maximum possible units between the supply and demand requirements. During allocation, the transportation cost is completely discarded (not taken into consideration).
2. Delete that row or column which has no values (fully exhausted) for supply or demand.
3. Now, with the new reduced table, again select the North-west corner cell and allocate the available values.
4. Repeat steps (2) and (3) until all the supply and demand values are zero.
5. Obtain the initial basic feasible solution.

Least Cost Method (LCM)

1. Allocating supply and demand requires choosing the least transportation cost cell that is present in the complete table.
2. Remove the exhausted row or column. No more allocation must take into account the deleted row or column.
3. Once more, pick the existing table's least cost column and distribute it. (Note: Choose the cells that allow for the most allocation if there are several least costs.)
4. Find the IBFS.

Vogel's Approximation Method (VAM)

1. By subtracting the lowest cost from the next-highest cost present in each row and column, penalties are determined for each row and column. The penalty is 0 if there are just the two lowest costs.

2. Choose the row or column with the biggest penalty and allocate data to the cell with the lowest cost in the chosen row or column. Choose the penalty when a row or column carries the minimum unit cost if there are two or more equal penalties. If a tie arises once again, choose the option that allows for the largest allocation.
3. The row or column that has met the demand and supply should be deleted.
4. Until all of the supply and demand are met, repeat steps (1) and (2).
5. Identify the IBFS.

Modified Distribution (MODI) Method

1. Select one of the three ways below to determine an IBFS:

    - North-West Conner Method (NWCM)
    - Least Cost Method (LCM)
    - Vogel's Approximation Method (VAM)

2. Using the formula $u_i + v_j = c_{ij}$, determine the values of the two variables u_i and v_j.
3. Use the formula $c_{ij} - (u_i + v_j)$ to calculate opportunity cost.
4. Verify each opportunity cost's indication. The suggested approach is the best one if the opportunity costs of all the vacant cells are either positive or zero. On the other hand, the provided method is not optimal, and additional transportation cost reductions are achievable if one or more unoccupied cells have a negative opportunity cost.
5. Decide which unoccupied cell in the following solution will have the lowest negative opportunity cost.
6. For the unoccupied cell chosen in the previous step, draw a closed route or loop. Please take note that only occupied cells and the initial unoccupied cell are allowed for the right-angle turn in this path.
7. Place a plus sign at the cell being assessed and the alternate plus and minus signs at the unoccupied cells on the corner points of the closed route.
8. Calculate the most units that can be delivered to this empty cell. The number of units that can be transported to the incoming cell is indicated by the smallest value with a negative position on the closed path. Now deduct this amount from all of the cells on the closed path's corner points that are marked with minus signs, and add it to all of the cells that are marked with plus signs. In this manner, an empty cell is converted into an occupied cell.

Stepping Stone Method

1. Select one of the following to determine an IBFS:
    - North-West Conner Method (NWCM)
    - Least Cost Method (LCM)
    - Vogel's Approximation Method (VAM)
2. In the case where m is the number of rows and n is the number of columns, make sure that the number of occupied cells is precisely equal to m + n - 1.
3. Decide on an empty cell. Trace a closed route starting from this cell and ending at the specified vacant cell, starting from this cell and ending at this cell.
4. On each corner cell of the recently traced closed route, alternately place plus (+) and minus (-) marks, starting with the plus sign at the unoccupied cell that will be used for evaluation.
5. Include the unit transportation costs for each cell that was identified along the closed path. This will result in a net change in cost.
6. Carry out steps 3 through 5 again to analyze every unoccupied cell.
7. Verify each net change in the unit transportation costs' indication. An optimal solution has been found if all the computed net changes are higher than or equal to zero. If not, go to step 8 since there are ways to enhance the present solution and lower the overall cost of transportation.
8. Most units that may be allocated to a cell must be chosen, and that cell must have the unoccupied cell with the greatest negative net cost change. The amount of units that can be transported to the entering cell is represented by the smallest value with a negative location on the closed path. This number must be added to the empty cell as well as to every other cell on the route denoted by a plus sign. On the closed path shown by a negative sign, subtract this number from the cells.

1.10. Objectives

The objectives of this research are mentioned below:

1) Proposed a novel approach to solve Minimum Spanning Tree (MST) using a Modified Ant Colony Optimization Algorithm.
2) Solve illustrative examples using the proposed method and compare results with Prim's and Kruskal's algorithms.
3) Apply the proposed algorithm to find the MST of the Indian Domestic Air Network.

4) Developed Ant Colony Optimization algorithm to find the best solution for both classical and Prohibited Route Transportation problems.
5) Solve examples of both types of TPs and compare results with other existing methods, such as North-West Conner Method (NWCM), Least Cost Method (LCM), Vogel's Approximation Method (VAM), and Optimal Solution.

CHAPTER 2

Literature Survey

Numerous network optimization research has been discovered using the network optimization model. The reduced complexity of time is one benefit of network optimization models. The Minimum Spanning Tree (MST) and Transportation Problem (TP) are the major topics of this research. Everything is constructed using the classical method and a modified ACO algorithm.

There are several approaches for finding the graph's MSTs. Based on how they are implemented, these algorithms are split into two groups, such as line-based (edge-based) and node-based (vertices-based). While Prim's and Dijkstra's algorithms are node-based, Kruskal's and Reverse Delete's algorithms are line-based. The first approach for finding a MST was created in 1926 by Czech researcher Otakar Borvka [1]. This algorithm is known as Borvka's algorithm. Then, Prim and Kruskal developed two different algorithms. One of the algorithms was developed by Robert C. Prim in 1957 after Vojtech Jarnk presented it in 1930. It was rediscovered in 1959 and given the name Prim's algorithm by Edsger W. Dijkstra [2]. In 1956, Joseph Kruskal [2] presented a different approach called Kruskal's algorithm. The Greedy Method sometimes referred to as the Reverse Deletes algorithm [3], is the opposite of Kruskal's algorithm for determining the shortest path.

H. N. Gabow, Z. Galil, T. Spencer, and R. E. Tarjan [4] proposed an efficient algorithm to find MST of both undirected and directed graphs in 1986. They developed fast techniques for finding MSTs in directed and undirected graphs using F-heaps and additional techniques. To find MST of an undirected graph they presented an $O(nlog\beta(m,n))$ – time algorithm and for directed graph $O(nlog\, n + m)$ – time algorithm.

In 2009, Stefan Ruzika and Horst W. Hamacher [5] surveyed multi-objective MST problems. They looked at research on MST issues involving two or more objective functions, each of which is a sum or bottleneck problem. Furthermore, existing methods are classified and described, and theoretical characteristics of various variants of this problem are discussed. This research contains a brief tabular summary of each of the publications that were evaluated.

D. Vijayalakshmir and R. Kalaivani [6] developed an algorithm to solve the Minimum Cost (Weight) Spanning Tree using Matrix Algorithm. They introduced a brand-new, straightforward method for finding an undirected connected weight graph's minimum cost-spanning tree. The use of MSTs was widespread. This algorithm is entirely based on an adjacency matrix of a weighted graph.

Based on the Vertex Associative Cycle Detection Method, proposed an Effective Greedy MST Algorithm by Prantik Biswasa, Mansi Goela, Harshita Negia, and Megha Dattaa [7] in 2016. To get an MST of an input-weighted undirected network, they describe a greedy approach based on cycle detection. The approach is based on the idea that a tree is a linked network with no cycles. The method chooses and determines if the highest degree vertex is a part of any cycle at each iteration to eliminate the costliest edge from the cycle connected to it. The iteration continues until all cycles are removed to produce the final MST. That method is straightforward, which makes it simpler to comprehend and use in any high-level language. The suggested method helps study a certain class of engineering and scientific challenges.

Murat Kirişci and Deniz Öncel [8] developed a new method involving MST for computer networks in a growing company in 2017. MSTs have been utilized in data mining, pattern identification, and machine learning for a long time. Traditional MST techniques have quadratic time complexity, making it difficult to apply them to big datasets. Their suggested algorithm is made to make this challenge easier.

In 2018, Sheikh Irfan Akbar and Shahid-ul-Islam [9] have provided descriptions of the two widely used algorithms (Prim's algorithm and Kruskal's algorithm) for resolving the MST problem. In addition, they discussed applications, time complexity, and a comparison of the two techniques. According to the survey, in dense networks with more edges than vertices, prim's method executes faster than Kruskal's approach.

In 2009, A. Khan, A. Aesha, and J. Sarker [10] suggested a novel algorithmic strategy for identifying MST. In their article, an alternative strategy was put out in which they made numerous uses of sets and disjoint sets union data structures to reduce the number of edges taken into account while calculating the MST of a graph.

A novel approach for discovering MSTs with undirected neutrosophic graphs was put out in 2019 by S. Dey, S. Broumi, and L.H. Son [11]. In their article, they talk about the MST issue of an undirected neutrosophic weighted linked graph where each arc's arc length is given a single-valued neutrosophic number rather than a real integer or fuzzy number. This sort of MST is described as a Neutrosophic Minimal Spanning tree (NMST). They discuss the use of neutrosophic numbers as arc lengths and how it is used for various MST issues that arise in the real world. Here, a brand-new method for creating a neutrosophic graph's MST is shown. The suggested approach uses the neutrosophic number as the arc length while using the uncertainty in the Kruskal algorithm for constructing MST. Different NMSTs are compared using a score function whose weights are calculated using the addition of neutrosophic numbers. They contrast the NMST's weight with a traditional MST's comparable weight using real values as the arc lengths. The suggested method is more effective than the NMST algorithms already in use since it allows for simple addition operations and the ranking of neutrosophic numbers. They used numerical examples to explain the suggested approach.

In 2017, P. Sreenivasulu Reddy and Abduselam Mahamed Derdar [12] developed the PS Reddy technique, which uses a matrix approach to compute the minimum and maximum spanning trees most straightforwardly for any weighted graph. However, for some graphs, the approach for determining the maximum spanning tree fails. By way of illustration, they provided proof.

Seth Pettie and Vijaya Ramachandran [13] proposed an optimal MST algorithm in 2002. They demonstrate that the decision-tree complexity and algorithmic complexity of the MST problem are equivalent. They specifically offer a deterministic technique that finds an MST of a network with n vertices and m edges in time $O(\tau^*(m, n))$, where τ^* is the minimal amount of edge-weight comparisons required to get the solution. The procedure is quite straightforward and is implementable on a pointer machine.

A Multi-objective Minimal Spanning Tree Problem (MMSTP) with uncertain issue parameters has been researched by Saibal Majumder [14] in 2022. The proposed uncertain multi-objective minimum spanning tree problem (UMMSTP) is then divided into two uncertain programming models, and the corresponding crisp equivalence models are researched. Finally, the problem is solved using a traditional multi-objective solution method called the epsilon-constraint method. Also used as solution approaches are two Multi-Objective Evolutionary Algorithms (MOEAs), Duplicate Elimination Non-dominated Sorting Evolutionary Algorithm (DENSEA) and Non-dominated Sorting Genetic Algorithm II (NSGAII). The practical issue of optimizing the distribution of petroleum products was resolved with the use of the suggested UMMSTP models, which involved looking for symmetry (balance) between the transportation cost and the transit time. After that, they examined the MOEAs' performance on five instances of the suggested problem that had been produced at random.

Paryati and Krit Salahddine [15] implement Kruskal's algorithm for MST in a graph in 2021. There are two ways to reach conclusions after testing and comparing Kruskal's method. strengths and weaknesses, for instance. When taking into account the strengths of Kruskal's independent algorithm, they discovered that the inclusion of weight sorting will make it easier to find the shortest path and will promote the construction of path tracks. Due to the need to sort thousands of vertices before creating the path, they discovered that if the number of vertices is very high, the process will be slower than Dijkstra's algorithm.

In 2008, A. Sayli and J. Alkhalissi [16] proposed a negligence MST Algorithm. In their study, they take into account a negligence algorithm to discover an MST differently. They compare their approach to Kruskal's MST algorithm in their study. To obtain the answers faster and more efficiently, they used certain graph instances.

Abhilasha R. [17] presented the Minimum Cost Spanning Tree utilizing Prim's method in 2013. The goal is to identify the minimum weight (cost) spanning the tree. They discovered that two separate sets of graph edges are used by Prim's method. If both the number of edges and the number of nodes are small, Prim's runs more quickly. For discovering the shortest path in a network, Prim's method functions similarly to Dijkstra's algorithm.

In 2020, a comparative analysis of the MST algorithm was proposed by Peace Ayegba, Joyce Ayoola, and Emmanuel Asani [18]. Implementing these algorithms and determining which of them is more effective is the aim of the study. To do this, they put the algorithms into practice and assess their complexity. The road network of Nigeria's 36 states was used to test both strategies, and the implementation's outcomes are described in their study.

Uthpala Ekanayake, Wasantha Daundasekara, and Pantalian Perera [2] proposed a novel approach for solving MST problems and TP using a modified ACO algorithm. In their analysis, they suggest a heuristic technique called the Modified Ant Colony Optimization Algorithm, which is based on the Ant Colony Optimization Algorithm (ACO) approach and has been shown to offer large-scale TPs with satisfactory near-optimal solutions. The transition Rule and Pheromone Update Rule were added to the ACO Algorithm in this unique technique to improve the degree of satisfaction with the optimal solution. Comparing the algorithmic technique suggested by their study to the well-known meta-heuristic algorithms in the literature, it is less complicated. Finally, they explain their findings and use a case study to demonstrate the suggested methodology.

The Ant Colony Optimization (ACO) Algorithm was proposed by Marco Dorigo [19, 20] in his 1992 doctoral thesis. It is a technique for figuring out the shortest route between two points. It is based on how ants move between their nest (colony) and a food source as they look for food. The approach they use makes it easier to choose the most efficient path between the source and the destination. The ACO Algorithm is a probabilistic method for resolving a variety of optimization issues. The MST issue, the shortest path problem, the traveling salesman problem, the traffic assignment problem, the scheduling problem, and so on are a few examples.

Among the many various kinds of transportation models that exist now, Hitchcock [21] first presented the most basic one in 1941. It was developed by Koopmans [22] in 1949 and Dantzig

[17] in 1951. Since then, several extensions to transportation models and approaches have been developed. The two solutions for TP are the Initial Basic Feasible Solution and the Optimal Solution (IBFS). The Least Cost Method (LCM) [23], the North-West Corner Method (NWCM) [23], and Vogel's approximation method (VAM) [23–25] are used to get IBFS, while the Optimal Solution is obtained using the MODI and the Stepping Stone Method.

In 2013, Abdul Quddoos, Shakeel Javaid, and M.M. Khalid [26] proposed a novel algorithm to solve bi-objective TP. Finding a non-degenerate compromise optimum solution for the Bi-Objective Transportation Problem (BOTP) is addressed in this study using a novel approach called the MMK method. To identify a nondegenerate compromise optimum solution to BOTP, the MMK technique uses the idea of the distance between two locations in the (x, y) coordinate. This method generates the set of all feasible non-degenerate efficient solutions. To demonstrate the proposed strategy, they provided a numerical example. The suggested approach and the already used methods have also been compared in their study.

A brief literature overview of TP with its mathematical models in balanced and unbalanced scenarios was provided in 2014 by Purnima Adhikari and Gyan Bahadur Thapa [27]. They provided a fundamentally workable solution and, as a result, the techniques to arrive at the balanced TP's optimal situation. Finally, they provided counterexamples for the primal-dual instance of the problem.

A. Edward Samuel [28] put out the Improved Zero Point Method (IZPM) for the TPs in 2012. In their paper, they discuss ways to improve Pandian and Natarajan's zero-point technique for TPs that are crisp or fuzzy. This straightforward rule enables us to develop the Improved Zero Point Method (IZPM) for the solution of Crisp and Fuzzy TPs. An optimal solution might come from such an adjustment. Sample problems were used to make a further comparison between the new method and the other transportation algorithms that are currently in use.

To identify an initial solution that is either the optimal solution or, at most, extremely near to it, Dhia Abdul Sattar Kadhem and Mardeen Shawkat Taher [29] devised an additional technique in 2013 employing the maximum "Supplies, Demands" and combining both of them with the minimum cost. The transportation algorithm follows the simplex method's exact stages. Instead of utilizing the standard simplex tableau, they utilize the transportation model's unique structure to arrange the calculation more practically. Their method considers the maximum supplies and the maximum demands and joins them with the related minimum available cost from the sources and the demands respectively to reach an initial solution that is closer to the optimal solution if it is not the North West Corner Method (NWCM) or Least Cost Method (LCM) or Vogel's Approximation Method (VAM) method.

In 2014, A. Arockiamary, and G.Pravina [30] proposed a method called "Application of Graph Theory to Find Shortest Path of TP". They examine how graph theory may produce TP by employing the shortest path in their work. They developed Dijkstra's algorithm and a practical solution to the problem of finding the shortest path between two points. Additionally, they created a network model of the TP for the situation, which was carefully examined to save shipping costs.

To reduce transportation costs, Mollah Mesbahuddin Ahmed, Aminur Rahman Khan, Faruque Ahmed, and Md. Sharif Uddin [31] proposed an effective method for finding an IBFS in 2016. Both the balanced and unbalanced TPs are subjected to methodical evaluation of the proposed approach. They demonstrate the generality of this approach by solving profit maximization TPs with it. Additionally, it was noted that the procedure typically produced minimal and maximal values for the TPs for cost minimization and profit maximization. Finally, the suggested approach claimed in this study asserts its broad applicability in resolving TPs.

Swati V. Kamble and Bhausaheb G. Kore [32] developed an algorithm called "A New Method to Obtain an IBFS of TP with the Average Opportunity Cost Method" in 2019. In their article, they provide a brand-new technique for obtaining IBFS for TP using the Average Opportunity Cost Method (AOCM). Calculations in terms of math and logic are extremely straightforward. They employ the Modified Distribution Approach (MODI) method to discover the IBFS and then improve it. Results showed that this strategy was more likely to produce an ideal solution for the majority of the TP than the other ways that were already in use. They provide excellent examples to show the same.

A modified approach based on LCM is presented in 2019 by A. K. M. Selim Reza, A. R. M. Jalal Uddin Jamali, and Bristy Biswas [33] for unbalanced TPs to determine an IBFS. Numerous numerical examples have been used to support the validity and efficacy of the suggested methods. To determine the effectiveness of the suggested algorithm in the case of unbalanced TP, illustration results were also contrasted with those obtained using existing methods, namely LCM and VAM. Additionally noted is the suitability of the modified algorithm for balanced TP. In the case of unbalanced TP, numerical results show that the modified algorithm is significantly more efficient than the alternatives.

In 2014, Md. Ashraful Babu, Md. Abu Helal, Mohammad Sazzad Hasan, and Utpal Kanti Das [34] introduced a new method called the "Implied Cost Method (ICM)," where feasible solutions are less expensive than VAM and often overlap with optimum solutions. However, they noted that while it is not guaranteed that ICM will always offer the most feasible solution, it frequently does.

The maximum difference method to find IBFS for TP was proposed by Smita Sood and Keerti Jain [35] in 2015. In this article, a novel approach to identifying a feasible initial basic solution to a transportation problem was put forward. The Maximum Difference Method was the name of the approach mentioned in this study. This approach provides a first, simple, and workable solution to the transportation problem, and it is frequently superior to Vogel's Approximation Method (VAM). The suggested approach was then shown using certain numerical examples.

In 2011, An Improved VAM for the TP was proposed by Serdar Korukoğlu and Serkan Balli [25]. One of the well-known transportation methods in the literature, VAM, was studied to find more effective initial solutions. It was suggested to modify VAM by including alternative allocation costs and total opportunity costs. To assess VAM and an improved version of VAM (IVAM), computational experiments were conducted. It was shown that for large-scale TPs, IVAM significantly obtains more effective initial solutions. The performance of IVAM compared to VAM was addressed in terms of the number of iterations and CPU time needed to arrive at the best solutions.

In 2018, T. Geetha and N. Anandhi [36] suggested a brand-new model addressing unbalanced TP. This approach is a methodical process that is simple to comprehend and use. They developed a new approach called the Standard Deviation Method (SDM) that, when compared to VAM, offers the best IBFS for unbalanced transportation situations. The decision-makers who are dealing with the unbalanced supply and demand will find this article quite useful in providing more possibilities. Finally, the suggested approach claimed in this study asserts its broad applicability in resolving TPs.

Using the approach referred to as "An Alternate Approach to finding an optimal Solution of a TP," Reena G. Patel, P.H. Bhathawala, and Bhavin S. Patel [37] suggested the concept in 2017 for identifying the IBFS as well as the optimal solution or close to the optimal solution of a TP. Here is an algorithm that focuses on empty cells before moving on. To further clarify the suggested algorithm, numerical examples were also given. However, the approach described above provides a step-by-step breakdown of the solution process for finding the best solution.

It was suggested in 2017 by Opara Jude, Oruh Ben Ifeanyichukwu, Iheagwara Andrew Ihuoma, and Esemokumo Perewarebo Akpos [38] that a fresh and effective method for identifying an IBFS to transportation issues be used. The "Inverse Coefficient of Variation Method (ICVM)" is the name of the suggested strategy, and seven numerical examples were used to demonstrate how it works. Six current approaches, including the Allocation Table Method (ATM), North West Corner Method (NWCM), Least Cost Method (LCM), Row Minimum Method (RAM), and Vogel's Approximation Method (VAM), were compared with the suggested strategy. The suggested Inverse Coefficient of Variation Method (ICVM) offers a better IBFS for every transportation issue considered in the experiment. Additionally, for many of the problems taken into consideration, the new approach provides the best answer.

In 2015, a method known as "TOCM-SUM" was proposed by A.R. Khan, A.Vilcu, N. Sultana, and S.S. Ahmed [39]. In their article, a novel heuristic for obtaining a TP's IBFS is presented. Numerous numerical examples were used to demonstrate the suggested strategy. A comparison of the new heuristic's results with those from the old heuristics demonstrates that the procedure described here produces better results. They created a powerful algorithm for TP cost minimization that is simple to grasp and produces better results in comparison to the approaches currently used and documented in the literature. Additionally, they supply MATLAB 7.7.0 code for the developed approach and demonstrate the validity of the code using several examples, demonstrating that the code yields the same result.

A novel method to determine the IBFS has been presented in 2019 by L. Kaur, M. Rakshit, and S. Singh [40]. It is based on allocating resources in zero-cost cells of a simplified transportation table. Comparing the suggested method to other ones already in use produces a solution that is close to optimal. The comparison between various current techniques was provided in their result analysis, and it was concluded that their new method yielded better IBFS than other existing methods, particularly for unbalanced TPs. The majority of the time provides an optimal or nearly optimal answer, requiring fewer iterations to obtain the optimal situation. However, it never guarantees optimality. As a result, after using their suggested strategy to find the answer, it is important to verify optimality. In addition, it is incredibly straightforward to use and understand.

To enhance the solution process and provide a better IBFS for the TPs in 2016, M. S. Uddin, A. R. Khan, C. G. Kibria, and I. Raeva [41] offered an effective improvement of LCM. An analysis of alternatives is also done to confirm the performance of the suggested strategy. Simulation results demonstrated the Improved Least Cost Method (iLCM). Finding the IBFS comes first in the TP solution process, followed by the necessary iterations to discover the best solution. In their investigation, iLCM seemed to be the most effective strategy in the majority of situations for finding IBFS in terms of proximity to the optimal situation. It would be reasonable to suppose that if the IBFS is the closest thing to the perfect answer, there won't be as many iterations after that. The optimization of transportation operations and reduction of transportation expenses as a result of this discovery are important for strengthening the market position of the company and saving time and resources. Solving 20 randomly selected numerical examples has been done in this article to demonstrate the utility of iLCM and to support its efficacy. It has been discovered that the approach works well for solving TPs. According to the simulation findings, the suggested solution, iLCM, ensures a minimal transportation cost while yet providing a high IBFS.

The fuzzy Goal Programming Method for Solving Multi-Objective TP was developed by K. Venkatasubbaiah, S. G. Acharyulu, and K. V. V. C. Mouli [42] in 2011. In this study, fuzzy membership functions were taken into account, along with deviation goals for each objective function. To demonstrate the viability of the suggested technique, the fuzzy max-min operator was used. To tackle the restricted optimization issue, the LINGO software program was employed. Two numerical cases were solved to demonstrate the suggested methodology, and the solutions' outcomes were compared with those of interactive, fuzzy, and deviation criterion techniques.

Z.A.M.S. Juman and his colleagues proposed a different solution TPs. An innovative method using row penalties to discover an IBFS to a TP was proposed in 2019 by Z.A.M.S. Juman and N.G.S.A. Nawarathne [24]. A numerical example was used to explain the new methodology. A comparative investigation of a collection of benchmark situations demonstrates that, except for one problem, the new technique offers the same IBFS or a better one. As a result, their novel approach might be seen as a different way to arrive at an IBFS than a TP. In 2021, he proposed a new algorithm for solving uncapacitated TP with interval-defined demands and supplier capacities [43]. Getting a homogenous product from several suppliers to multiple customers while reducing transportation costs is the goal of the uncapacitated TP. In reality, the needs and capacities may not be understood precisely but change over time. Therefore, for each potential solution, the actual values of all parameters will determine the ultimate transport cost. The decision-makers aim to reduce transportation expenses and are concerned with the calculation of the worst-case scenario. A few randomly generated numerical example problems and benchmark examples were used to test the performance of these novel strategies. In 2022, he published another article called "Close Interval Approximation of Pentagonal Fuzzy Numbers for Interval Data-Based TPs" [44]. In the Interval Data-based Transportation Problem (IDTP), where supply, demand, and cost are all expressed as interval numbers, solutions are presented. The closed interval approximation of pentagonal fuzzy numbers, one of the best interval approximations, is suggested for solving the IDTP. First, the pentagonal fuzzification approach transforms the IDTP into a fuzzy TP in the suggested closed interval approximation method (Method-1). Two novel ranking techniques are then introduced to convert the pentagonal fuzzy number into the appropriate crisp (non-fuzzy) value, based on centroid and in-center triangle notions. Following that, Vogel's approximation approach and the modified distribution method were combined to get the best result. Through benchmark examples and novel cases, the suggested Technique-1 outperforms the previously described and proposed Method-2 according to a recent method.

Several research papers on TPs were published by E.M.U.S.B. Ekanayake and his colleagues. In 2021, They suggested a different approach that identifies the optimum or nearly optimum solution to the TP. This approach, which uses an iterative technique, applies to both balanced and unbalanced TPs. Additionally, it should be noted that, in comparison to other approaches now in use, this method takes fewer iterations to obtain optimality. Additionally, they have created a brand new technique for finding the best solution to TPs that are both balanced and unbalanced [45]. He uses a bipartite graph to analyze the best way to solve the TP in 2021. Topological spaces, graphs, and TPs are all parts of this process. It first turns the TP into a graphical presentation before changing into a different graphical picture. Then, following the suggested algorithmic approach, they were able to determine the most affordable way to carry items from one set of supply vertices to another [46]. Additionally, he introduced a Modified Ant Colony Optimization Algorithm (MACOA) in 2020 to achieve an IBFS to a TP. The suggested approach is uncomplicated, easy to use, and provides approximately optimal results in a few iterations. Solving validity and applicability instances have also been used to support the efficiency of this technique. To determine the potential significance of their MACOA, a thorough numerical investigation is conducted. In terms of the quality of the solution, the comparison analysis shows that both the MACOA and the current JHM are effective when compared to the methodologies examined in this work. However, in reality, because JHM calculation takes a long time, researchers and practitioners recommend using their suggested MACOA when dealing with big-sized TPs [47]. Moreover, based on the "max-flow," they created a different technique for determining the maximum flow between a network's source and destination nodes. Additionally, the new maximum flow algorithm serves as the foundation for a novel algorithmic method for solving the TP (minimizing the transportation cost). Also, it should be noted that this approach only needs a limited number of repetitions to reach its full potential. The algorithmic methodology used in this work is simpler than the well-known meta-heuristic algorithms in the literature [48]. In 2022, he wrote a paper with the title "An Improved Ant Colony Algorithm to Solve Prohibited Transportation Problems". To obtain an IBFS to a prohibitive transportation problem (PTP), this research developed a heuristic approach and an enhanced ant colony optimization technique. On the other hand, employing the PTP in the suggested technique results in the best IBFS for a prohibited transportation problem and outperforms existing methods with less computational complexity and time. As a result, the suggested methodologies provide a compelling substitute for conventional problem-solving techniques. In some numerical instances, the suggested method's feasible solution and optimum solution are the same [49]. Furthermore, in 2022, they examined several TPs and mathematical model types. A high-level review of numerous transportation-related problems and mathematical models is provided in that study paper. This may be used to successfully address a range of business issues relevant to the distribution of goods, often known as TPs. This article gives a quick introduction to a few different forms of TPs, such as classical TPs, TPs with mixed constraints, interval TPs, bottleneck TPs, multi-objective TPs, and fuzzy TPs. That chapter also included the research on the TP that has been done so far by several scientists and statisticians, as well as mathematical models for each [50].

E.M.D.B. Ekanayake [51, 52] discovers a novel approach algorithm for computing an IBFS for TPs in 2022. In this research work, a model technique based on the average unit cost value of columns and rows is suggested to determine the IBFS for TPs with balanced and unbalanced components. To address the problem by balancing supply and demand, a new table is built utilizing the unit cost values and average unit cost values in the columns and rows and comparing them to the TP. The suggested process is seamless, simple to comprehend, and straightforward to use. Compared to well-known meta-heuristic algorithms in the literature, the algorithmic method proposed in this study is less difficult. Lastly, present a case study to illustrate the suggested strategy.

Joseph Ackora-Prah [53] proposed an algorithm called "A Two-Phase Method for Solving Transportation Models with Prohibited Routes" in 2022. This study proposes a two-phase approach to solve a TP with prohibited routes. In the first stage, using Floyd's technique, an All-Pairs Least Cost Problem (APLCP) is used to assign a non-discretionary penalty cost of $M_{ij}^* \leq M$ to each of the n current prohibited routes. In step two, the original issue is replaced with the new penalty values in the appropriate places, and the resulting model is then solved using the transportation method. The findings indicate that rationally presenting this adjusted penalty cost (M*) is a suitable solution. As a result, using the M at your discretion does not ensure that your model solutions are valid.

Moreover, Walid Kamal Abdelbasset [54], A. Seethalakshmy [55], S. Krishna Prabha [56], Mollah Mesbahuddin Ahmed [31], K.P.O.Niluminda [56 – 59], Krishna Prabha [61], Sharmistha Jana [62], Haleemah Jawad Kadhim [63], Abul Kalam Azad [64], Md. Salehin Ferdous Kader [65], D. Santhosh Kumar [66], Bimal Sinha [67], P. K. Giri [68], Mohammad Asim Nomani [69], Khilendra Singh [70], Lushu Li [71], K. Bharathi [72], Lakhveer Kaur [73], Rakesh Verma [74], Doke D.M. [75], Sanjay R. Ahir [76], Kavita [77], Ekanayake E.M.U.S.B.

[78], Khilendra Singh [79], M. Afwat A.E. [80], Yadvendra Kacher [81], Khilendra Singh [82], Abouzar Sheikhi [83], T. Karthy [84], M. Afwat A.E. [85], M. Zangiabadi [86], Mohamed Sadeg Zayed Ogbi [87], Kirti Kumar Jain [88], Osuji, George A. [89], M. A. Mujeeb Khan [90] were proposed several algorithms to solve different types of transportation problems.

CHAPTER 3

Methodology

3.1. The Modified Ant Colony Optimization Algorithm (MACOA)

A population-based search method for determining the ideal path that is based on ant behavior is known as an Ant Colony Optimization (ACO) algorithm. Ants move from their nest (let's say, a city i) to a food source using pheromones. The probabilistic transition rule and pheromone update rule are applicable if an ant k is currently located in city i and wishes to relocate to city j. More details of the probabilistic transition rule and pheromone update rule are mentioned in sub-section 1.7.1 and 1.7.2.

The probabilistic transition rule of the Ant Colony Optimization (ACO) Algorithm was used for this development. The pheromone trail is used by an ant to move from one source to another. As a result, if i^{th} ant decides to travel next unvisited j^{th} destination (city) is given by the following formula (6).

We assume $\alpha = 0$ and $\beta = 0.5$

$$P_{ij} = \begin{cases} \dfrac{\left[{}^{1}/_{w_{ij}}\right]^{0.5}}{\sum_{z \in Allowed_k}\left[{}^{1}/_{w_{ij}}\right]^{0.5}} & ; i^{th}\ ant\ visits\ the\ j^{th}\ city \\ 0 & ; Otherwise \end{cases} \tag{6}$$

Here,

P_{ij} = Probability to branch from i^{th} node to j^{th} node.

w_{ij} = Cost or distance between i^{th} node and j^{th} node

The steps for solving the minimum spanning problem are as follows.

3.2. Ant Colony Optimization for Minimum Spanning Tree Algorithm

Step 1: Calculate the path based on the probability of every edge by using the Modified Ant Colony Optimization Algorithm.

Step 2: Determine the degree of each vertex.

Step 3: Select the vertices with the highest degree that are not adjacent to each other.

Step 4: Determine the maximum probability path and choose the best two paths from each selected vertex, deleting the other probability paths.

Step 5: If the degree of the vertex is below 3, proceed to step 6. Otherwise, proceed to step 4 for every vertex.

Step 6: If the subgraphs are disconnected, connect every subgraph by using the maximum probability of two paths.

Step 7: If the graph has cycles, remove those cycles by deleting the minimum probability paths.

Step 8: If the termination condition is chosen, the best vertices with the best paths are chosen.

In the first step, the algorithm calculates the probability of each edge is included in the minimum spanning tree by using the modified ant colony optimization algorithm. This algorithm works by simulating the behavior of a colony of ants that search for food in a graph. By assigning probabilities to each edge based on the amount of pheromone deposited by the ants, the algorithm can identify the most promising paths between vertices. In step 2, the algorithm determines the degree of each vertex in the graph, i.e., the number of edges that connect to that vertex. This information is used in later steps to select the most promising vertices to add to the minimum spanning tree. In step 3, the algorithm selects the vertices with the highest degree that are not already connected by an edge in the minimum spanning tree. This ensures that the minimum spanning tree remains connected and avoids forming cycles. In step 4, the algorithm identifies the maximum probability path between each selected vertex and its neighbors and chooses the best two paths to add to the minimum spanning tree. The other probability paths are deleted to avoid forming cycles.

The flow chart representation of the newly proposed method is also provided in Figure 3.1.

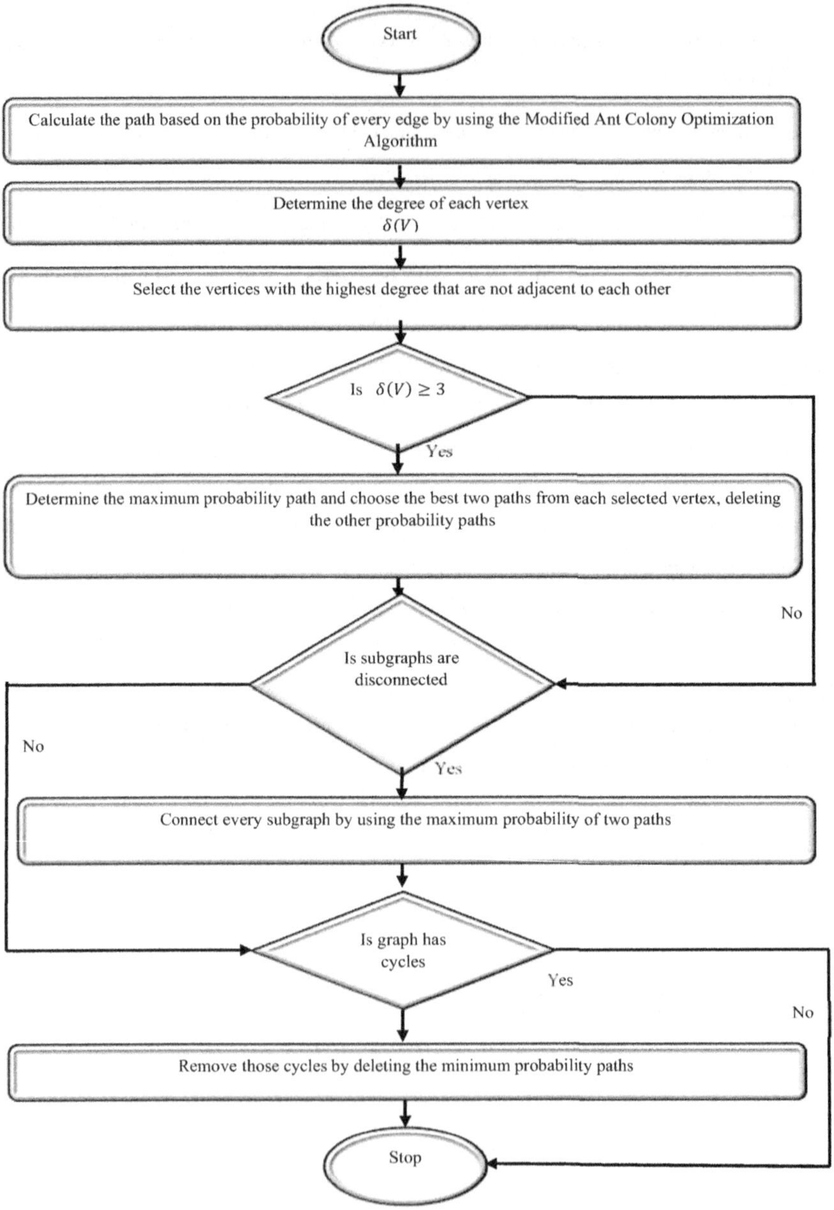

Figure 3.1: Flow chart representation of the new algorithm, MACOA

In step 5, the algorithm checks if the degree of the selected vertices is below 3. If so, it proceeds to step 6. Otherwise, it repeats step 4 for every vertex to add more edges to the minimum spanning tree. In step 6, the algorithm checks if there are any disconnected subgraphs in the minimum spanning tree. If so, it connects them by adding edges between the subgraphs using the maximum probability of two paths. In step 7, the algorithm checks if there are any cycles in the minimum spanning tree. If so, it removes them by deleting the minimum probability paths in the cycles. In the final step, the algorithm checks if the termination condition has been reached, such as a maximum number of iterations or a desired level of solution quality. If so, it selects the best vertices and paths in the minimum spanning tree as the final solution.

Overall, your proposed algorithm combines the strengths of the Modified Ant Colony Optimization Algorithm and traditional minimum spanning tree algorithms to efficiently find high-quality solutions to this important problem in computer science and operations research.

3.3. MST of Indian Domestic Air Network

The organization in charge of managing India's domestic and international airports is known as the Airport Authority of India. The 137 airports that the Airports Authority of India (AAI) oversees include 24 international airports, 10 customs airports, and 103 domestic airports.

In this research, 73 airports are considered. Figures 3.2 and 3.3 show the map of the Indian Domestic Air Network. Each airport has a unique code. The table of names and codes of airports is represented in Appendix A.

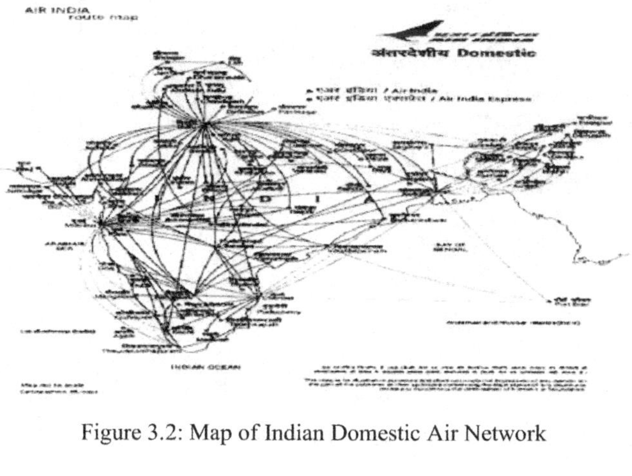

Figure 3.2: Map of Indian Domestic Air Network

The proposed Modified Ant Colony Algorithm will apply to find the MST of the Indian Domestic Air Network. There are 73 airports and 140 airlines in this study. Here, we assume airports as nodes (vertices) and airlines as a link (edges).

3.3.1. Data Collection

The distances and exact locations of airports in India were obtained from online free software: https://distancecalculator.globefeed.com/India_Distance_Calculator.asp. The Domestic Air Network in India graph was plotted using Microsoft Office 2016.

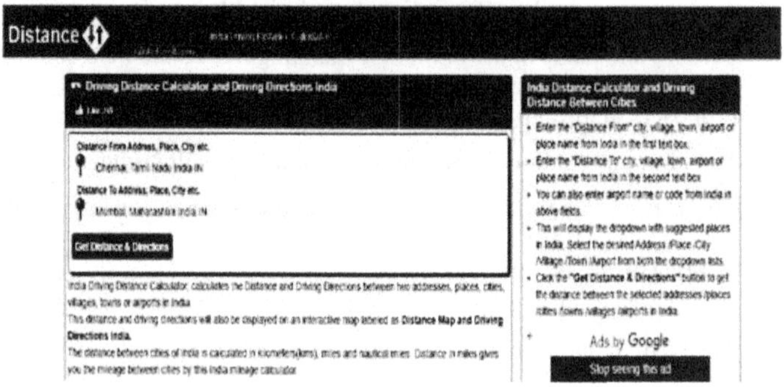

Figure 3.4: Interface of Distance Calculator Software

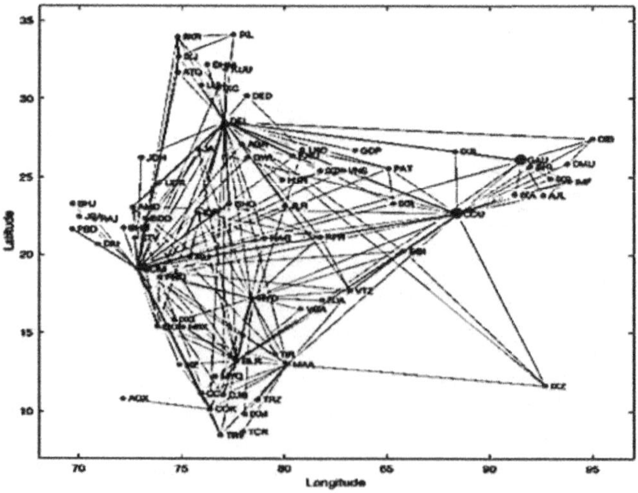

Figure 3.3: Indian Domestic Air Network with Airport Code

3.3.2. Data Analysis

The data obtained from the above software was analyzed using Microsoft Excel 2016. Using this software calculated every path based on probability from i^{th} node to j^{th} a node using Modified Ant Colony Optimization Equation shows in equation no 6.

3.4. Proposed New Algorithm to Solve both Classical and Prohibited Route TPs

This section describes the proposed new algorithm for TPs. Both balanced, unbalanced classical (acceptable), and prohibited route TPs can solve using this proposed method. The optimal or near-optimal solution can be obtained using this technique. The following is a description of the steps in the suggested method:

Step 1: Check whether your TP is balanced or not. If it is not balanced introduce a new dummy row or column at the end of the source rows or destination columns.

Step 2: Construct the Transportation graphical representation.

Step 3: Find the Minimum Spanning Tree (MST) of the graphical network using any MST algorithm.

Step 4: Arrange the weight of MST in ascending order. If MST has an edge with zero weight, ignore that edge and go to the next edge in order.

Step 5: Assign the corresponding allocated value (x_{ij}) to cells as lined up.

Step 6: Adjust the supply and demand of the selected cell in TT and cross out the satisfied row or column.

Step 7: If the selected cell is already crossed out, then go to the next maximum value of the order.

Step 8: After assigning the every x_{ij} value to cells as lined up, if all the supplies and demands are not satisfied in TT, then assign the x_{ij} value to the cell which has minimum cost value (c_{ij}).

Step 9: After all, requirements are satisfied calculated the total transportation cost by the following equation:

$i.e.$

$$\sum_{i=1}^{m}\sum_{j=1}^{n} x_{ij}c_{ij}$$

Here,

c_{ij} = Unit transportation cost

x_{ij} = Corresponding allocated value.

CHAPTER 4

Results

Operations research uses graph theory in a variety of ways. A collection of points connected by a finite number of lines is referred to as a graph. The proposed method for finding the minimum spanning tree of an undirected graph will be illustrated using two separate instances in this section. Also, the proposed method will apply to the Indian Domestic Air Network to find the MST of that network. Using that we can demonstrate how airports connect with the minimum number of air routes and the total distance of that network. Finally proposed MST algorithm will develop to find the IBFS of any both balance and unbalanced classical and prohibited route TPs.

4.1 Proof of the proposed method for minimum spanning trees examples

This section discusses the performance of the new MST approach.

4.1.1 Minimum Spanning Tree Example 1 [91]:

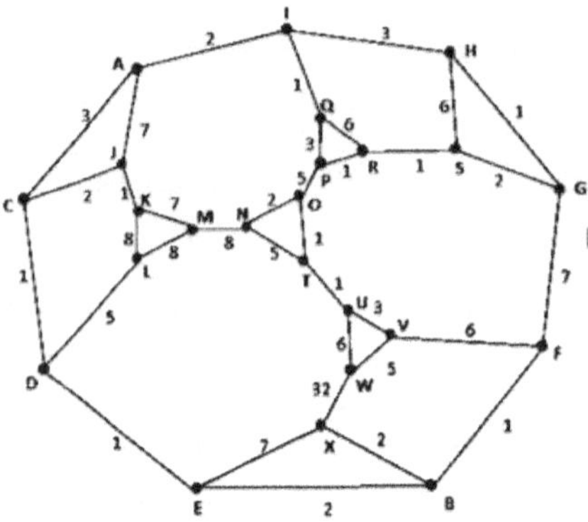

Figure 4.1: Connected Undirected Graph 1

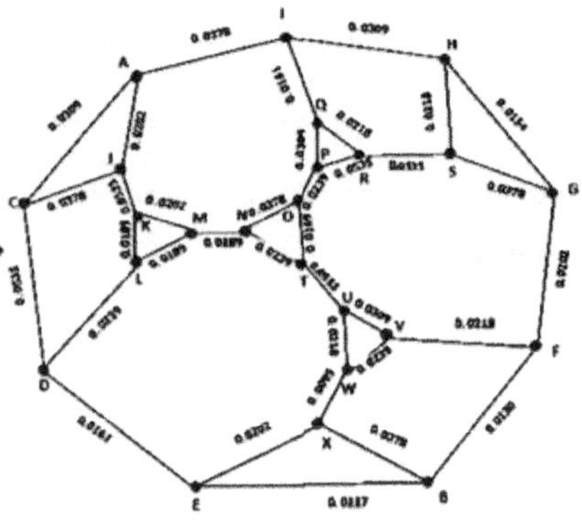

Figure 4.2: Graph 1 with Probabilities

Step 2: Determine the degree of each vertex.

Step 3: Select the vertices with the highest degree that are not adjacent to each other.

Step 4: Determine the maximum probability path and choose the best two paths from each selected vertex, deleting the other probability paths.

Step 5: If the degree of the vertex is below 3, proceed to step 6. Otherwise, proceed to step 4 for every vertex.

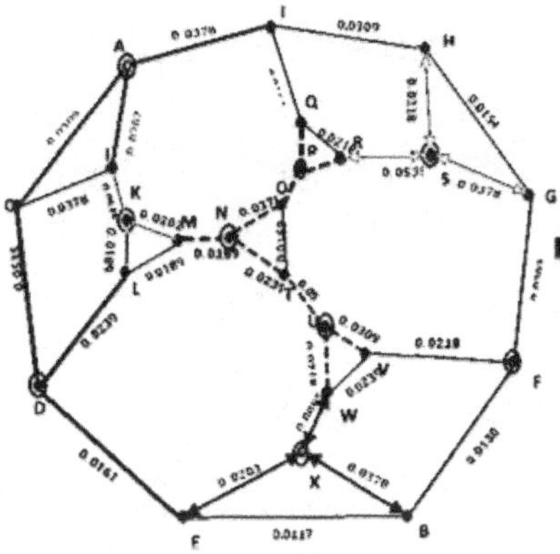

Figure 4.3: Reduced Graph 1.1

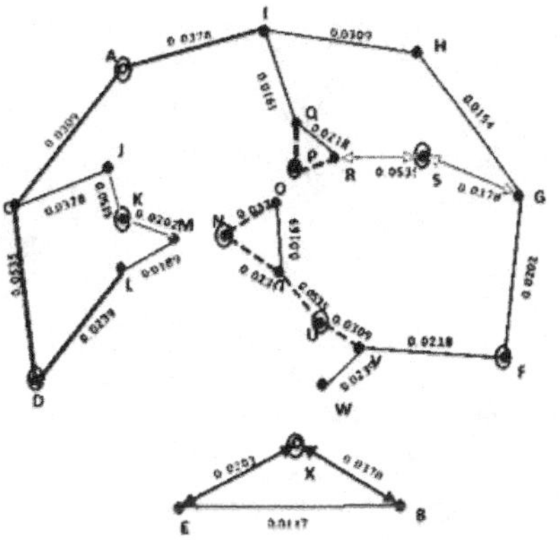

Figure 4.4: Reduced Graph 1.2

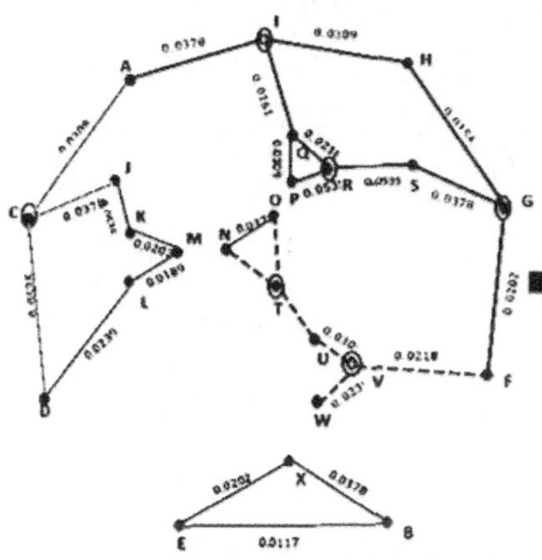

Figure 4.5: Reduced Graph 1.3

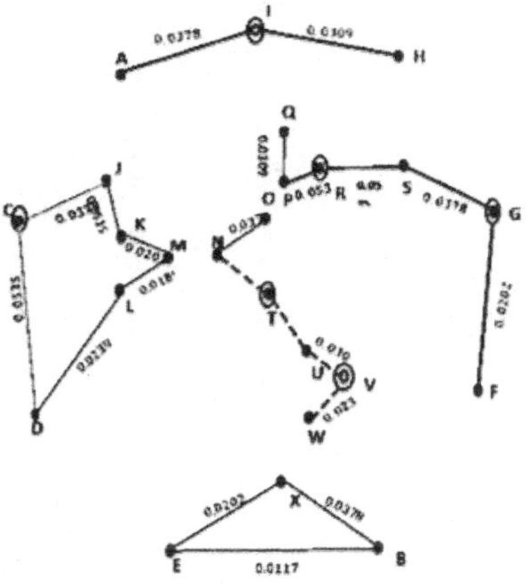

Figure 4.6: Reduced Graph 1.4

Step 6: If the subgraphs are disconnected, connect every subgraph by using the maximum probability of two paths.

Step 7: If the graph has cycles, remove those cycles by deleting the minimum probability paths.

Figure 4.7: Reduced Graph 1.5

Figure 4.8: MST with probabilities

The Minimum Spanning Tree (MST)

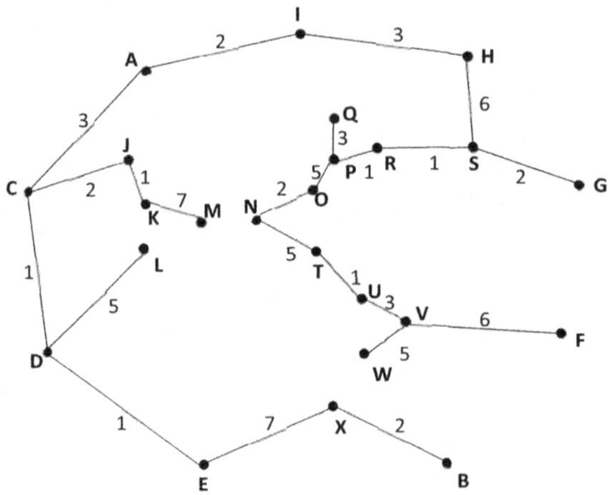

Figure 4.9: The Minimum Spanning Tree of Graph 1

This connected graph has 24 nodes (vertices) and 36 edges. The minimum spanning tree has 23 edges and the total weight of the minimum spanning tree (MST) is 84 units. It is compression is in Section 5.1.

4.1.2 Minimum Spanning Tree Example 2 [92]:

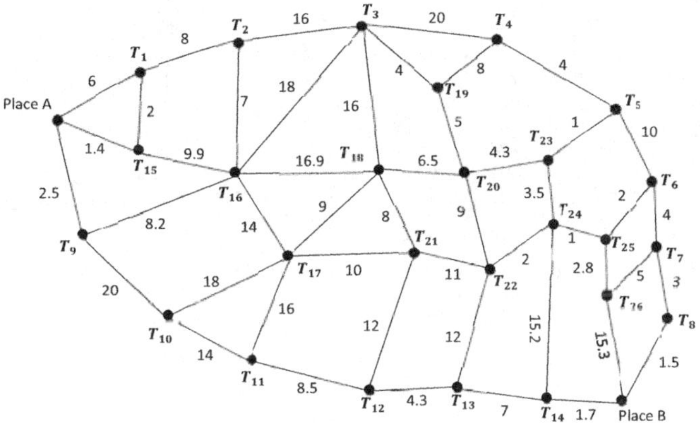

Figure 4.10: Undirected Connected Graph 2

Step 1:

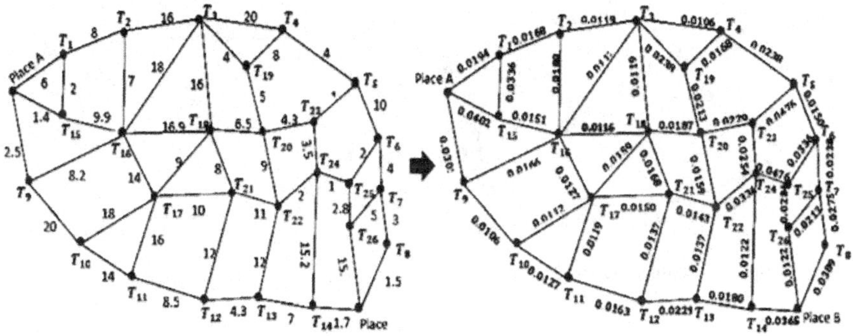

Figure 4.11: Graph 2 with Probabilities

Steps 2, 3 & 4:

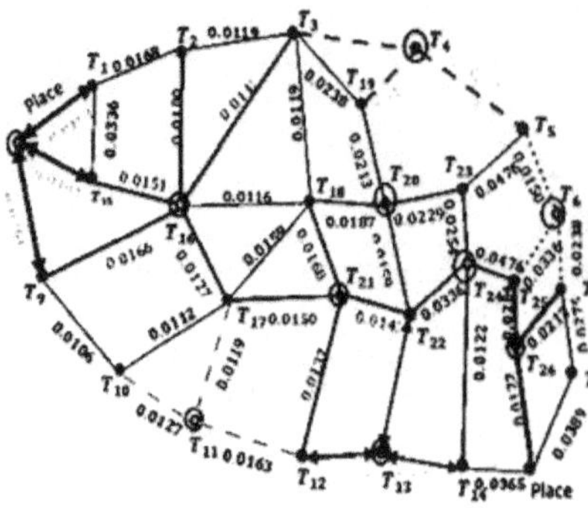

Figure 4.12: Reduced Graph 2.1

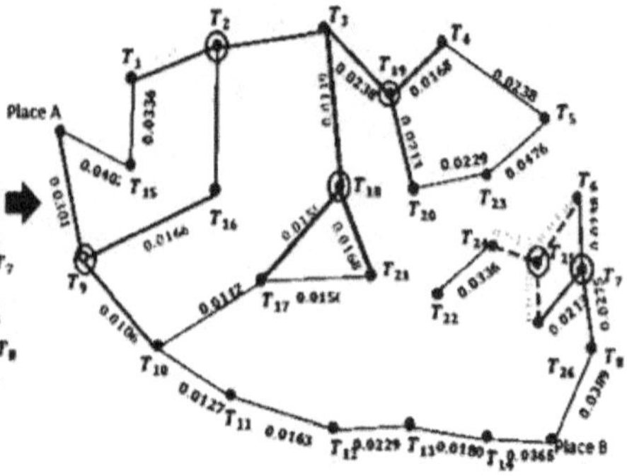

Figure 4.13: Reduced Graph 2.2

Step 6:

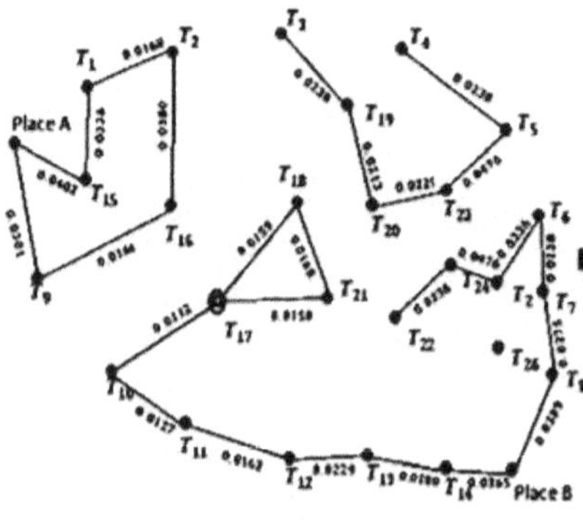

Figure 4.14: Reduced Graph 2.3

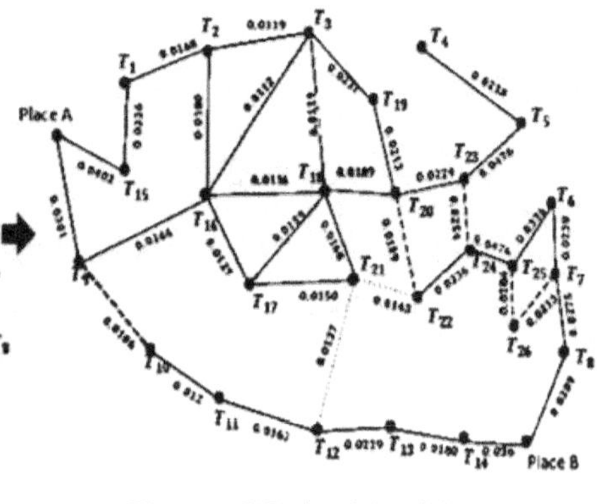

Figure 4.15: Reduced Graph 2.4

Step 7:

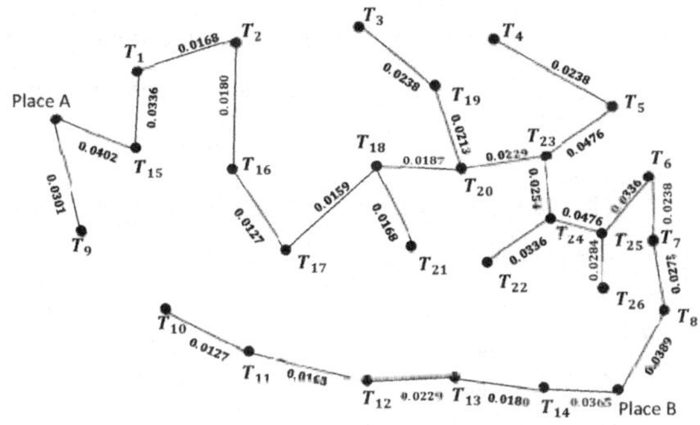

Figure 4.16: MST with Probabilities

The Minimum Spanning Tree (MST)

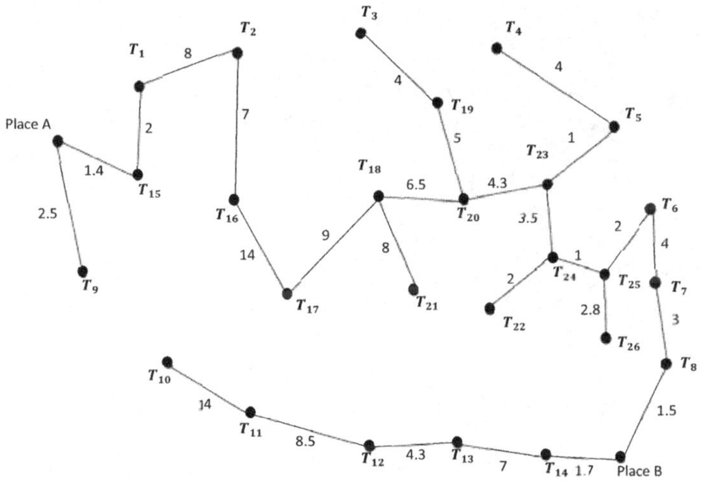

Figure 4.17: The Minimum Spanning Tree of Graph 2

This connected graph has 28 nodes (vertices) and 48 edges. The minimum spanning tree has 27 edges and the total weight of the minimum spanning tree (MST) is 132 units. It is compression is in Section 5.1.

4.2 Application of MST: Indian Domestic Air Network (IDAN)

In this section, the proposed MST method will apply to the Indian Domestic Air Network. There are 73 airports and 140 airlines in this network. We take airports as nodes or vertices and airlines as lines or edges. The name and unique codes of airports are represented in Appendix A. The map of IDAN shows in Figures 3.2 and 3.3. The distances between airports (in kilometers) show in Appendix B.

Step 1: Calculate the path based on the probability of every vertex by using the Modified Ant Colony Optimization Algorithm.

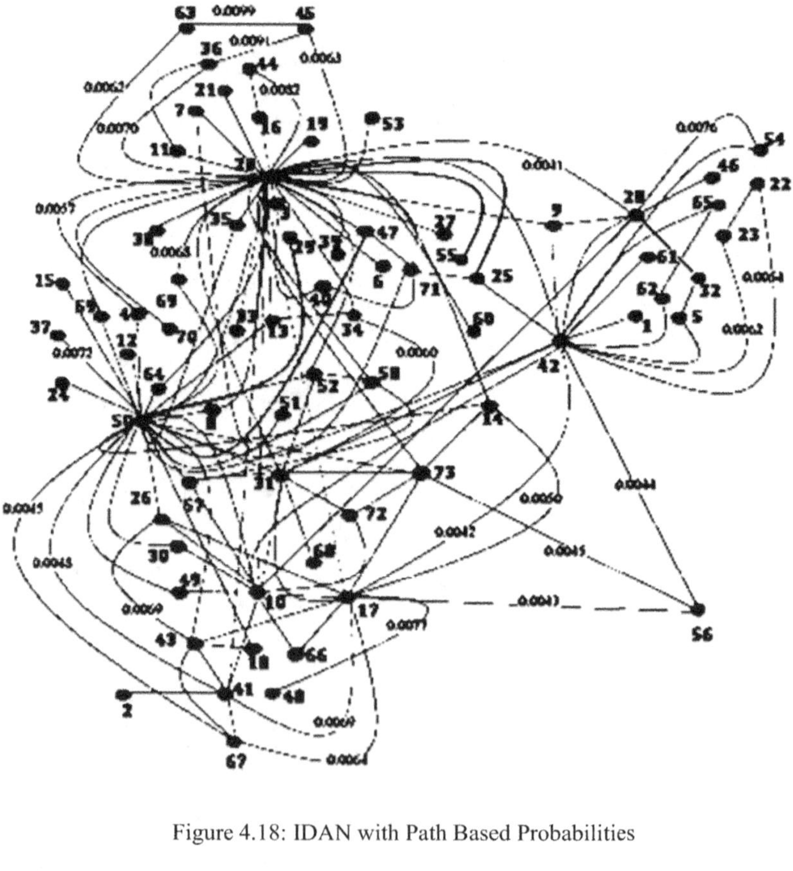

Figure 4.18: IDAN with Path Based Probabilities

Steps 2 – 5:

The red-circled vertices are the highest-degree vertices which not adjacent to each other.

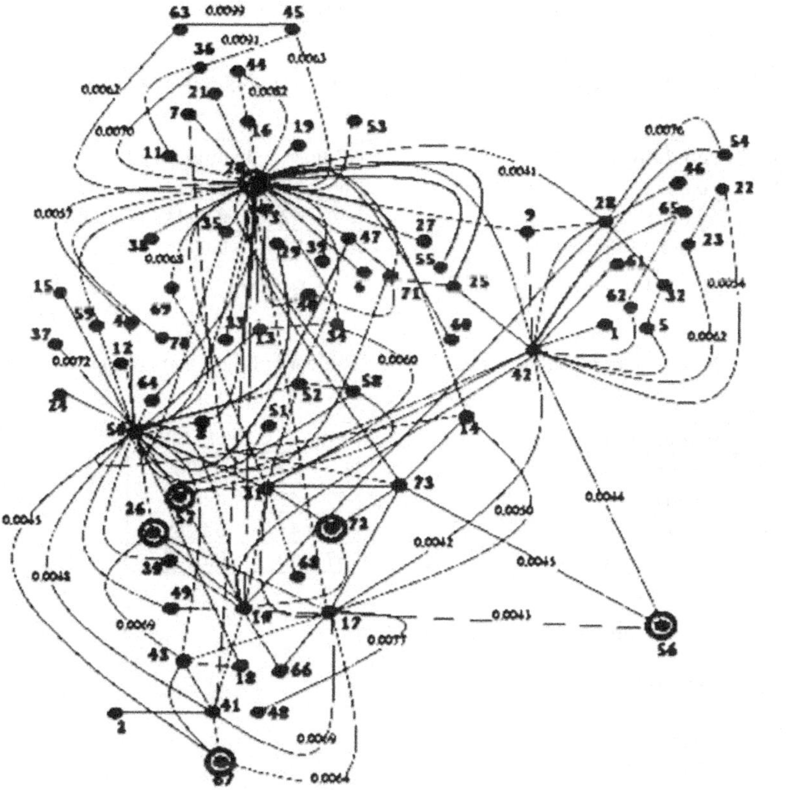

Figure 4.19: After Selecting the Highest Degree Vertices of the Graph of IDAN_1

Table 4.1: Edges that are joined to vertex 20

Edges	Probability	Edges	Probability	Edges	Probability
20 – 4	0.0057	20 – 73	0.0043	20 – 9	0.0047
20 – 70	0.0056	**20 – 3**	**0.0118**	20 – 28	0.0041
20 – 38	0.0072	20 – 34	0.0061	20 – 63	0.0062
20 – 59	0.0051	20 – 39	0.0079	20 – 36	0.0070
20 – 69	0.0068	20 – 52	0.0054	20 – 11	0.0092
20 – 64	0.0052	20 – 58	0.0051	20 – 7	0.0078
20 – 35	0.0104	20 – 6	0.0066	20 – 21	0.0078
20 – 50	0.0047	20 – 47	0.0077	20 – 16	0.0103
20 – 10	0.0038	20 – 71	0.0061	20 – 44	0.0082
20 – 8	0.0050	20 – 14	0.0044	**20 – 19**	**0.0109**
20 – 33	0.0061	20 – 27	0.0062	20 – 45	0.0063

20 – 41	0.0035
20 – 13	0.0065
20 – 31	0.0044
20 – 17	0.0038

20 – 60	0.0050
20 – 55	0.0054
20 – 25	0.0053
20 – 42	0.0044

20 – 53	0.0103

Table 4.2: Edges that are joined to vertex 26

Edges	Probability
26 – 10	**0.0072**
26 – 17	0.0058
26 – 43	0.0069
26 – 50	**0.0077**

Table 4.3: Edges that are joined to vertex 56

Edges	Probability
56 – 17	0.0043
56 – 42	**0.0044**
56 – 73	**0.0045**

Table 4.4: Edges that are joined to vertex 57

Edges	Probability
57 – 10	0.0059
57 – 31	**0.0071**
57 – 50	**0.0142**
57 – 58	0.0054

Table 4.5: Edges that are joined to vertex 67

Edges	Probability
67 – 17	0.0064
67 – 41	**0.0113**
67 – 43	**0.0089**
67 – 50	0.0045

Table 4.6: Edges that are joined to vertex 72

Edges	Probability
72 – 10	0.0071
72 – 31	**0.0097**
72 – 73	**0.0093**

The yellow color cells are the paths that have the highest probabilities. You should select those paths and delete other paths adjacent to corresponding vertices.

After doing step 4 to Figure 4.19, the resulting graph shows in Figure 4.20 graph.

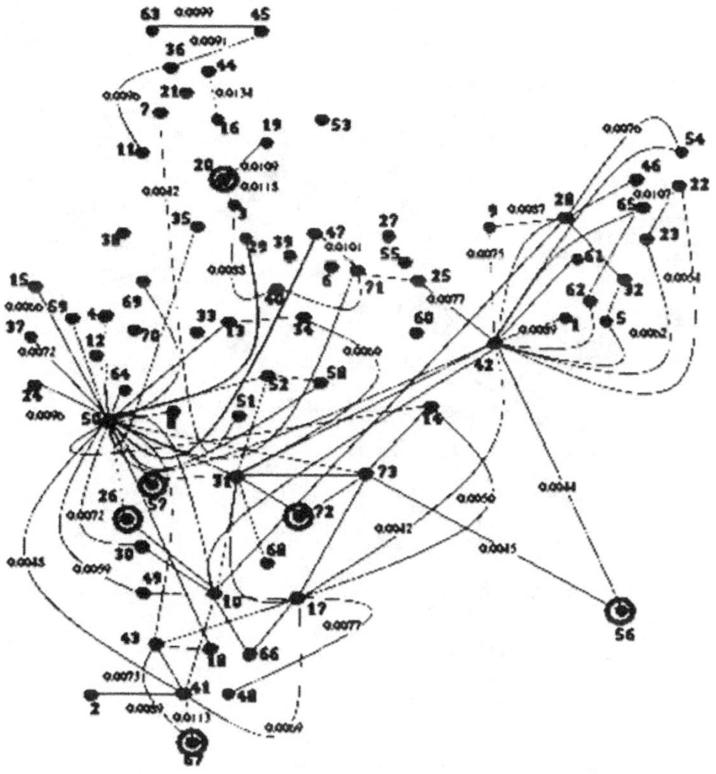

Figure 4.20: Reduced graph of IDAN_1

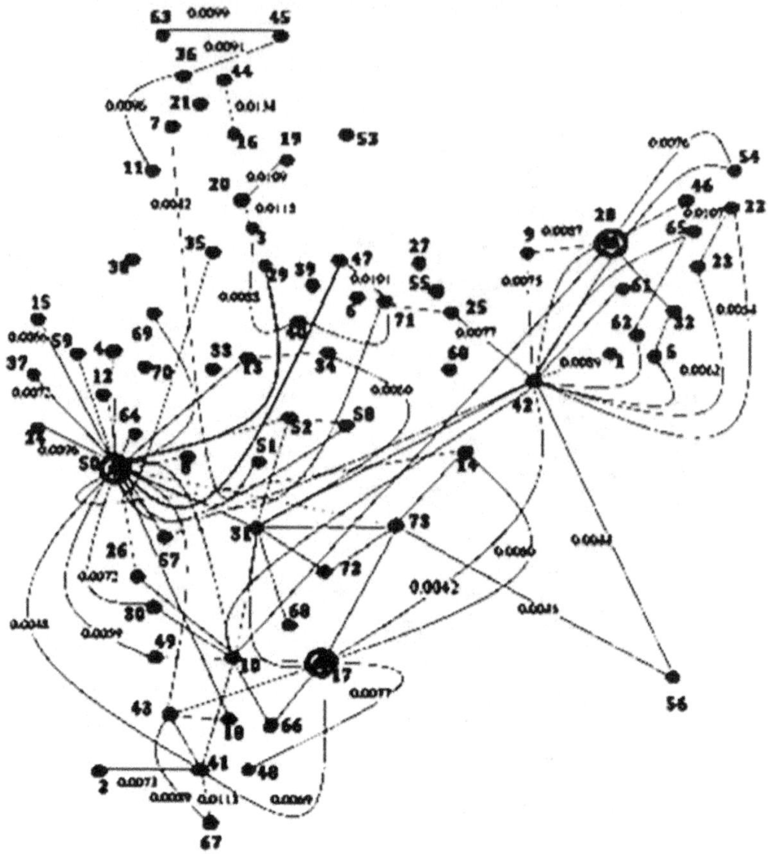

Figure 4.21: After Selecting the Highest Degree Vertices of the Graph of IDAN_2

Table 4.7: Edges that are joined to vertex 50

Edges	Probability	Edges	Probability	Edges	Probability
50 – 24	0.0096	50 – 29	0.0051	50 – 10	0.0055
50 – 37	0.0072	50 – 47	0.0046	50 – 18	0.0050
50 – 15	0.0066	50 – 71	0.0045	50 – 41	0.0048
50 – 59	0.0077	50 – 42	0.0039	50 – 43	0.0051
50 – 12	0.0090	50 – 52	0.0060	50 – 26	0.0077
50 – 4	0.0075	50 – 8	0.0095	50 – 30	0.0072
50 – 64	**0.0105**	50 – 58	0.0051	50 – 49	0.0059
50 – 69	0.0063	50 – 14	0.0043	50 – 67	0.0045
50 – 35	0.0052	50 – 73	0.0048		
50 – 13	0.0062	**50 – 57**	**0.0142**		

Table 4.8: Edges that are joined to vertex 17

Edges	Probability
17 – 10	**0.0096**
17 – 14	0.0050
17 – 31	0.0070
17 – 41	0.0069
17 – 42	0.0042
17 – 43	0.0070
17 – 48	0.0077
17 – 66	**0.0092**
17 – 73	0.0063

Table 4.9: Edges that are joined to vertex 28

Edges	Probability
28 – 9	0.0087
28 – 10	0.0035
28 – 32	**0.0095**
28 – 42	0.0071
28 – 46	**0.0094**
28 – 54	0.0076

After doing step 4 to Figure 4.21, the resulting graph shows in Figure 4.22 graph.

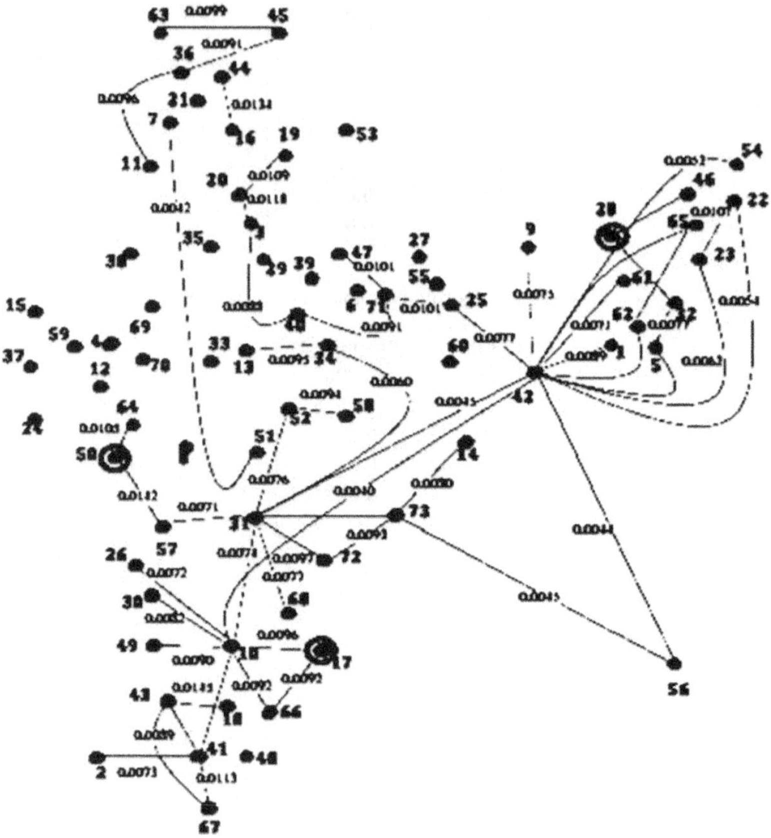

Figure 4.22: Reduced graph of IDAN_2

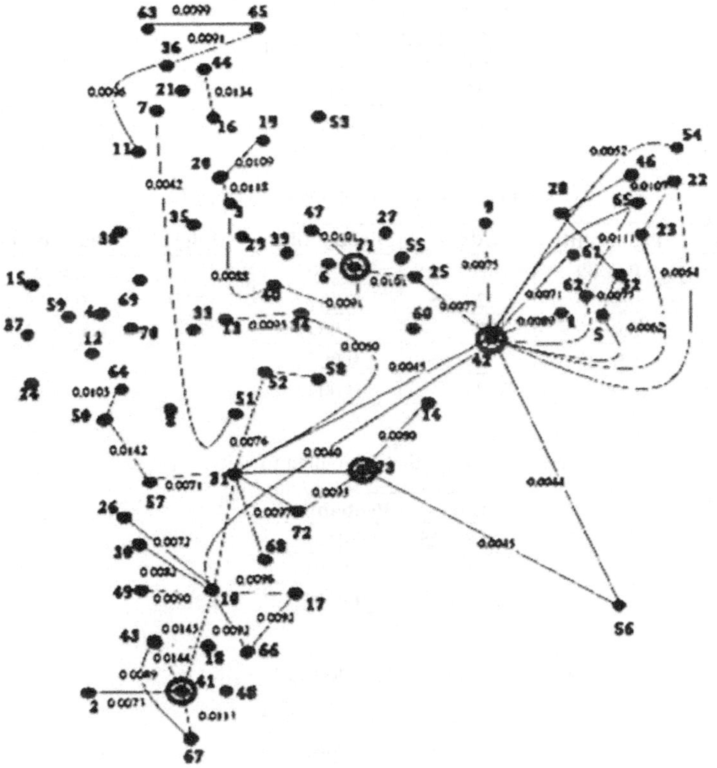

Figure 4.23: After Selecting the Highest Degree Vertices of the Graph of IDAN_3

Table 4.10: Edges that are joined to vertex 41

Edges	Probability
41 – 2	0.0073
41 – 10	0.0082
41 – 43	**0.0144**
41 – 67	**0.0113**

Table 4.11: Edges that are joined to vertex 42

Edges	Probability	Edges	Probability	Edges	Probability
42 – 1	**0.0089**	42 – 23	0.0062	42 – 61	0.0071
42 – 5	0.0075	**42 – 25**	**0.0077**	42 – 62	0.0069
42 – 9	0.0075	42 – 31	0.0045	42 – 65	0.0063
42 – 10	0.0040	42 – 54	0.0052		
42 – 22	0.0054	42 – 56	0.0044		

Table 4.12: Edges that are joined to vertex 71

Edges	Probability
71 – 25	**0.0105**
71 – 40	0.0091
71 – 47	**0.0101**

Table 4.13: Edges that are joined to vertex 73

Edges	Probability
73 – 14	**0.0080**
73 – 31	0.0070
73 – 56	0.0045
73 – 72	**0.0093**

After doing steps 4 to Figure 4.23, the resulting graph is shown in Figure 4.24 graph.

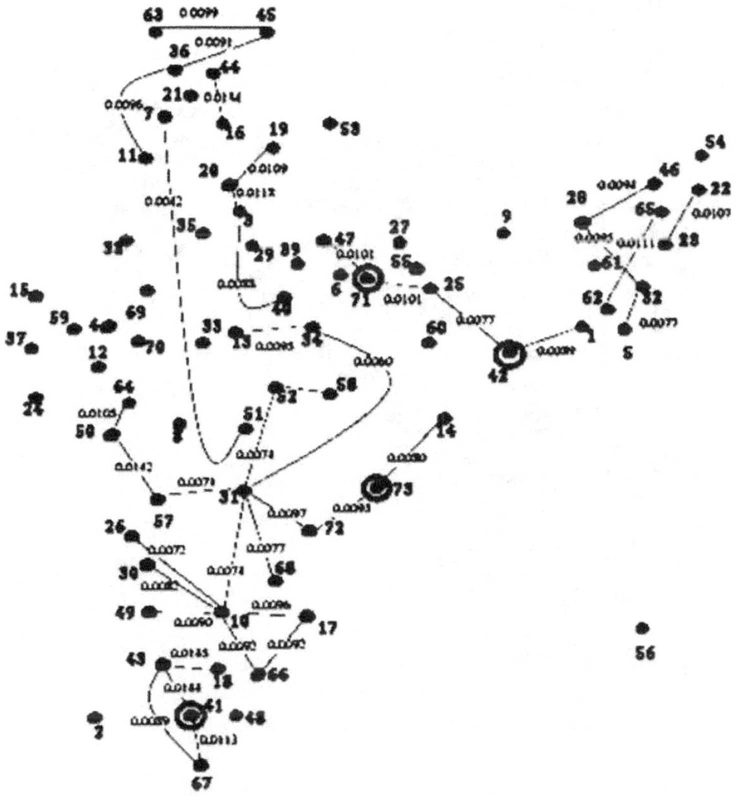

Figure 4.24: Reduced graph of IDAN_3

Figure 4.25: After Selecting the Highest Degree Vertices of the Graph of IDAN_4

Table 4.14: Edges that are joined to vertex 31

Edges	Probability
31 – 10	0.0074
31 – 34	0.0060
31 – 52	0.0076
31 – 57	0.0071
31 – 68	**0.0077**
31 – 72	**0.0097**

After doing step 4 to Figure 4.25, the resulting graph shows in Figure 4.26 graph.

Figure 4.26: Reduced graph of IDAN_4

Figure 4.27: After Selecting the Highest Degree Vertices of the Graph of IDAN_5

Table 4.15: Edges that are joined to vertex 10

Edges	Probability
10 – 17	**0.0096**
10 – 26	0.0072
10 – 30	0.0082
10 – 49	0.0090
10 – 66	**0.0092**

Table 4.16: Edges that are joined to vertex 43

Edges	Probability
43 – 18	**0.0145**
43 – 41	**0.0144**
43 – 67	0.0089

After doing step 4 to Figure 4.27, the resulting graph shows in Figure 4.28 graph.

Figure 4.28: Reduced graph of IDAN_5

Step 6: If the subgraphs are disconnected, connect every subgraph by using the maximum probability of two paths.

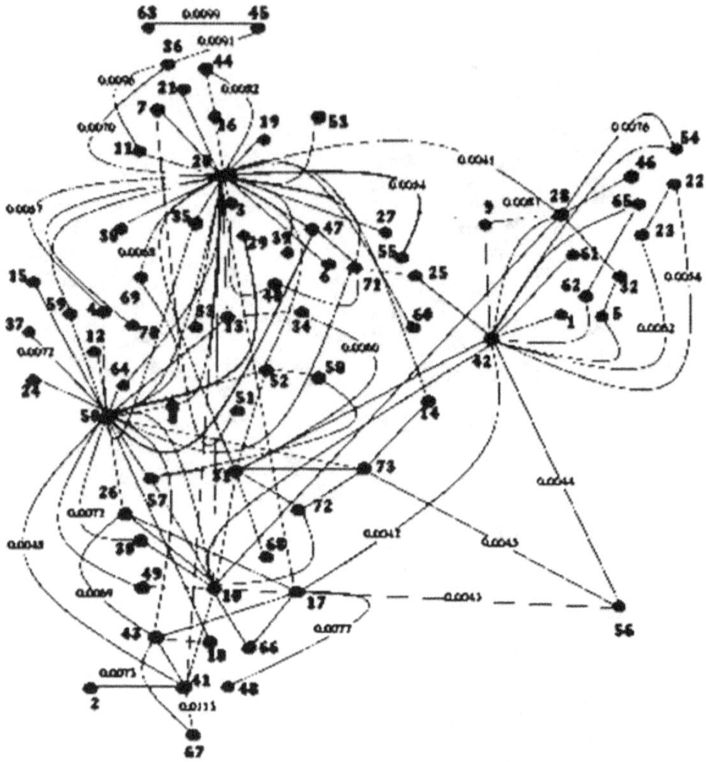

Figure 4.29: Connect every subgraph by using the maximum probability of two paths

Step 7: If the graph has cycles, remove those cycles by deleting the minimum probability paths.

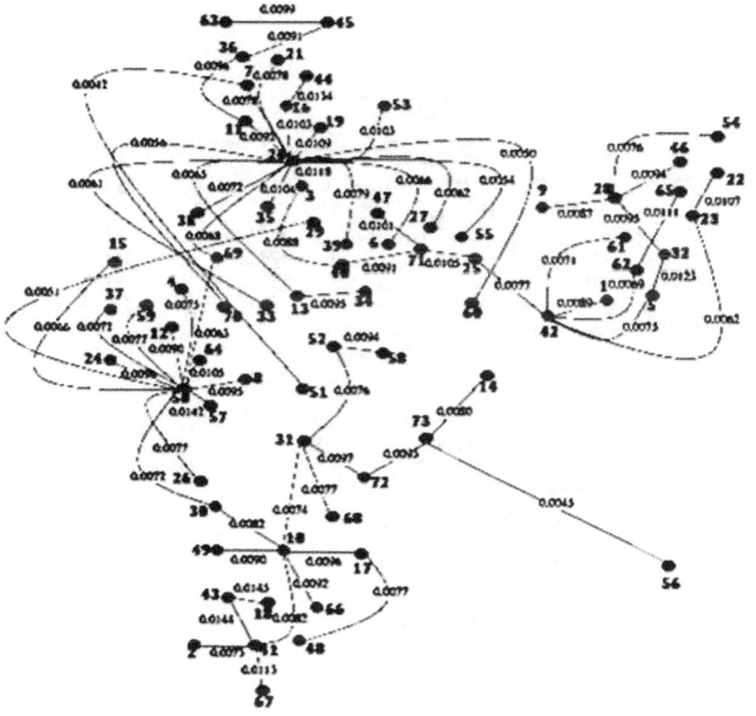

Figure 4.30: Remove the cycles by deleting the minimum probability paths

The Minimum Spanning Tree of Indian Domestic Air Network

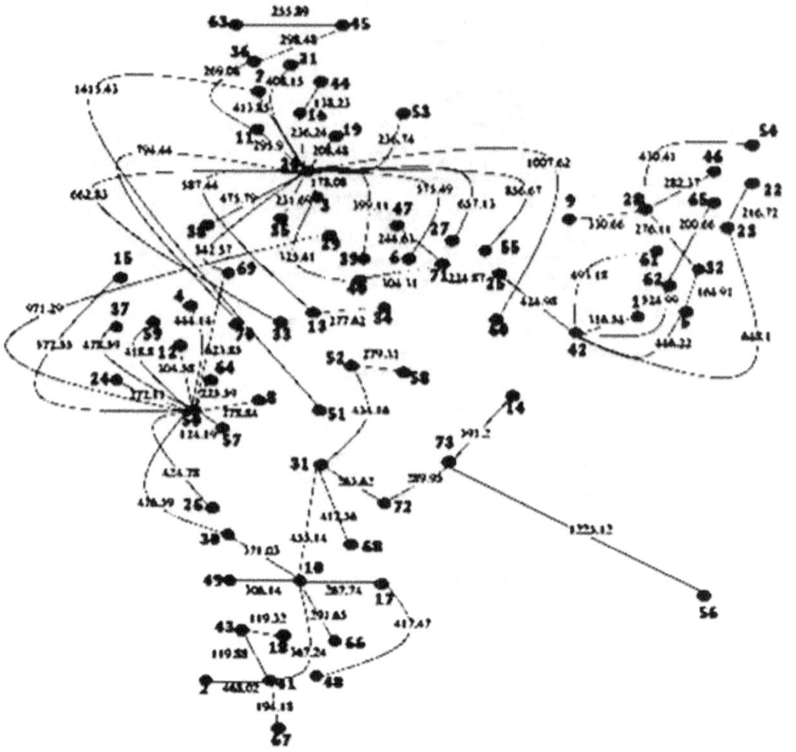

Figure 4.31: MST of IDAN Using Proposed Method

The total weight of the Indian Domestic Air Network's Minimum Spanning Tree using the proposed method = 29572.65 km. It is compression is in Section 5.2.

4.2.1 MST of Indian Domestic Air Network Using Prim's Algorithm

In this part, find the MST of IDAN using Prim's algorithm. The steps of Prim's algorithm are mentioned in sub-topic 1.5.

Figure 4.32: Prim's Algorithm Graph_1

Figure 4.33: Prim’s Algorithm Graph_2

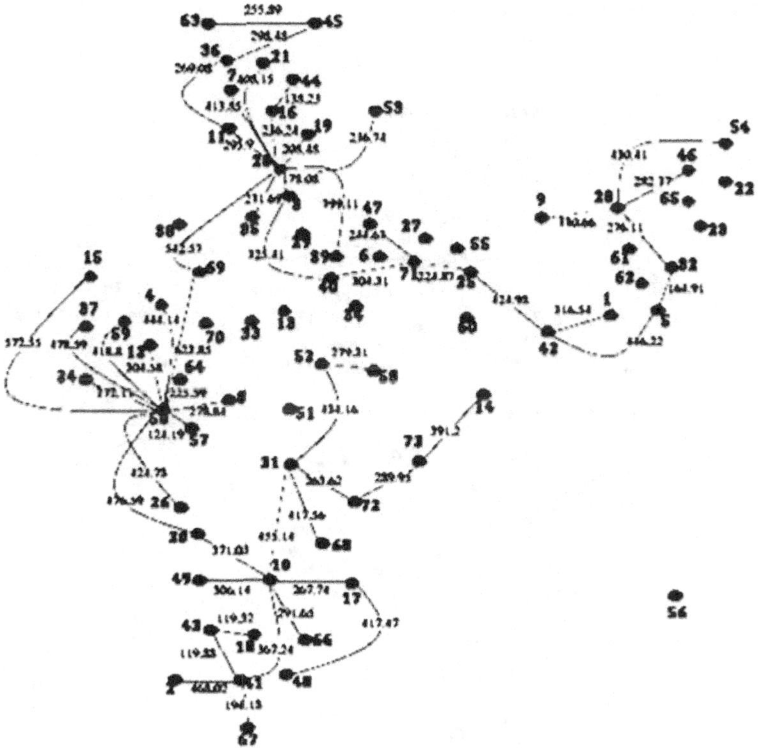

Figure 4.34: Prim’s Algorithm Graph_3

The MST of IDAN using Prim's algorithm

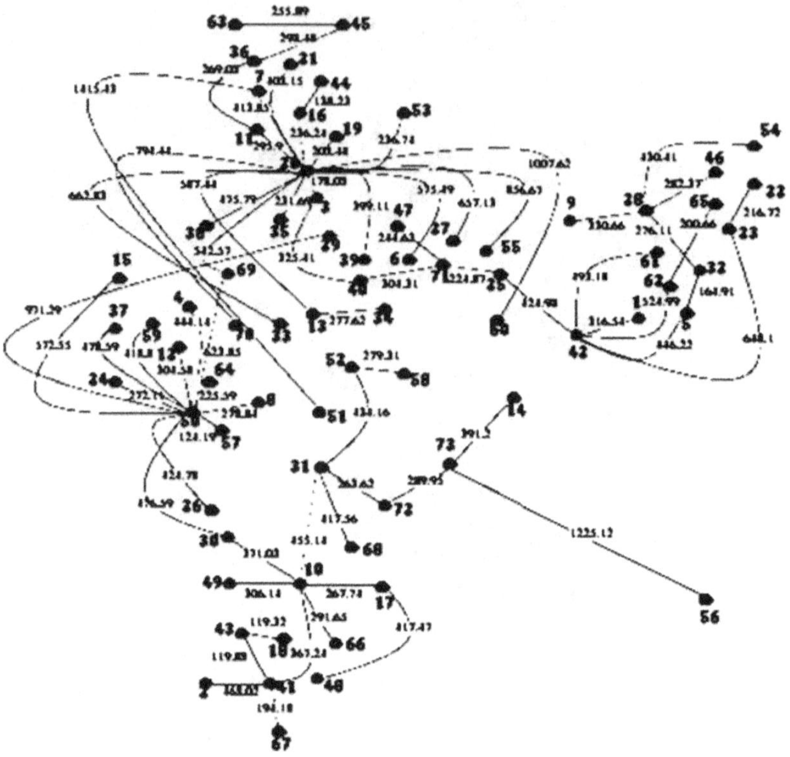

Figure 4.35: MST of IDAN Using Prim's Algorithm

The total weight of Indian Domestic Air Network's Minimum Spanning Tree using Prim's algorithm = 29572.65 km

4.2.2 MST of Indian Domestic Air Network Using Kruskal's Algorithm

In this part, find the MST of IDAN using Kruskal's algorithm. The steps of Kruskal's algorithm are mentioned under sub-topic 1.6.

✓ Edge weight less than 300 km:

Figure 4.36: Kruskal's Algorithm Graph with Edge Weight Less than 300km

✓ Edge weight less than 500 km:

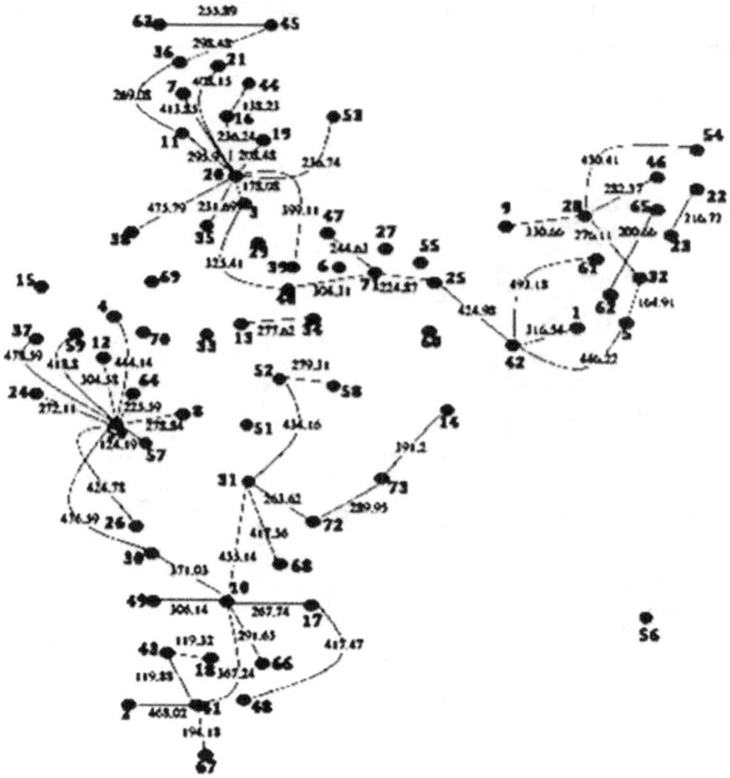

Figure 4.37: Kruskal's Algorithm Graph with Edge Weight Less than 500km

✓ Edge weight less than 700 km:

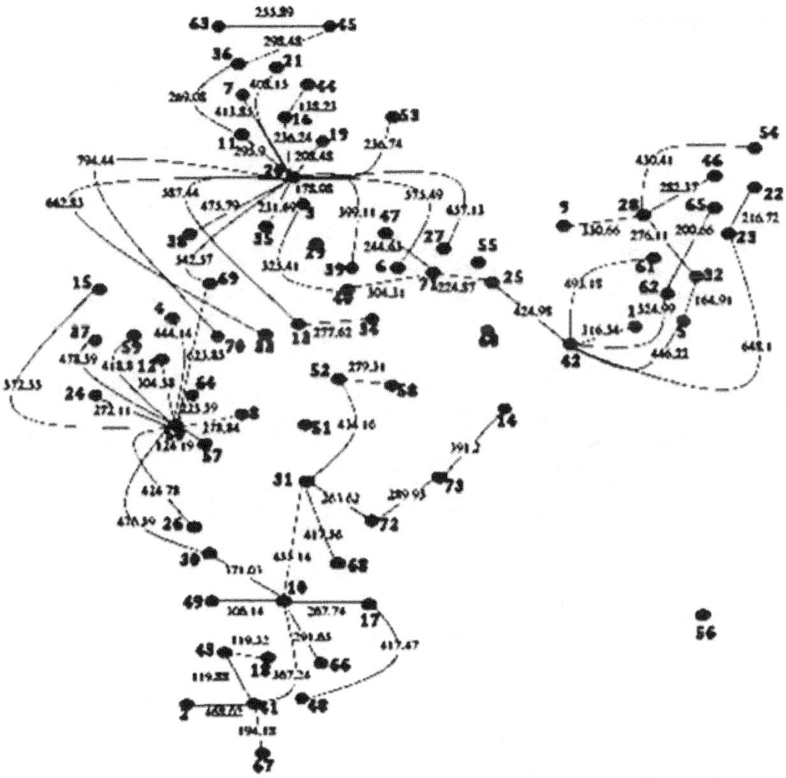

Figure 4.38: Kruskal's Algorithm Graph with Edge Weight Less than 700km

✓ Edge weight less than 2050 km:

The Minimum Spanning Tree

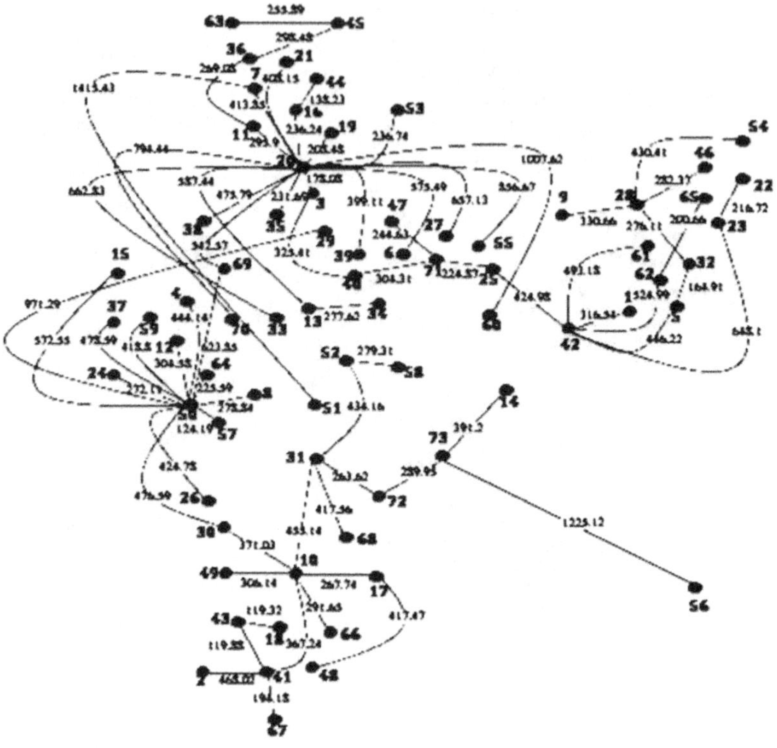

Figure 4.39: MST of IDAN Using Kruskal's Algorithm

The total weight of Indian Domestic Air Network's Minimum Spanning Tree using Kruskal's algorithm = 29572.65 km

4.3 Illustrative Examples of Transportation Problems

The TP is one of the most colorful and intriguing problems in Operation Research (OR). Many researchers have focused their efforts on resolving the problem using various approaches. Certain approaches concentrated on determining the IFS, whereas others focused on determining the OS. As a result, this study considered this fallback and proposed a method known as "classical transportation problems" (CTP) as well as "prohibited route transportation problems" (PRTP).

4.3.1. Example 1: Balanced Classical Transportation Problem_1 (BCTP_1) [46]

Step 1:

Table 4.17: Initial Balanced Classical Transportation Problem (BCTP_1)

	D_1	D_2	D_3	D_4	Supply
S_1	50	60	100	50	20
S_2	80	40	70	50	38
S_3	90	70	30	50	16
Demand	10	18	22	24	

Step 2:

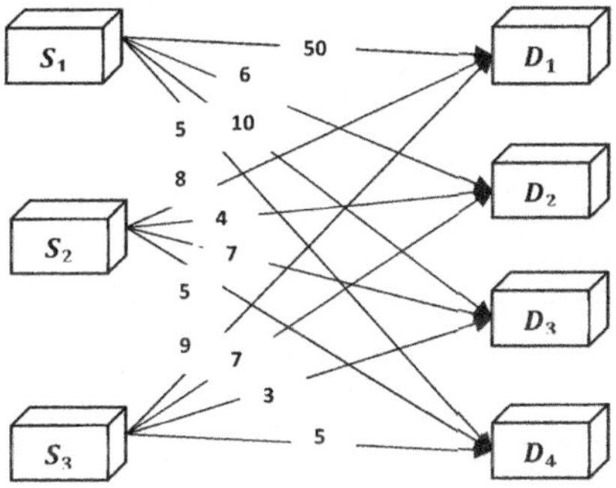

Figure 4.40: Transportation Network of BCTP_1

Step 3:

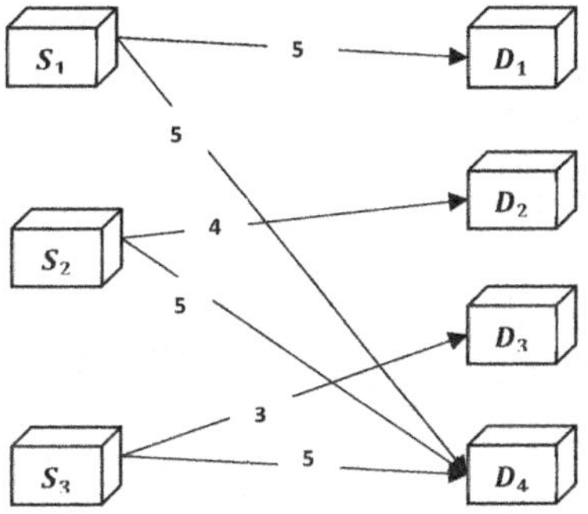

Figure 4.41: Minimum Spanning Tree of Transportation Network of

Step 4: Allocation Order: S_3D_3, S_2D_2, S_1D_1, S_1D_4, S_2D_4

Steps 5 – 8:

Table 4.18: Allocation table of BCTP_1

	D_1	D_2	D_3	D_4	Supply
S_1	50 (10)	60	100	50 (10)	~~20~~,~~10~~,0
S_2	80	40 (18)	70 (6)	50 (14)	~~38~~,~~20~~,~~6~~,0
S_3	90	70	30 (16)	50	~~16~~,0
Demand	~~10~~,0	~~18~~,0	~~22~~,~~6~~,0	~~24~~,~~14~~,0	

Step 9:

Total transportation cost = (50 × 10) + (50 × 10) + (40 × 18) + (70 × 6) + (50 × 14) + (30 × 16) = 3320

The compression results of this example are in Section 5.3.1.

4.3.2. Example 2: Unbalanced Classical Transportation Problem_1 (UCTP_1) [46]

Step 1:

Table 4.19: Initial Unbalanced Classical Transportation Problem (UCTP_1)

	D_1	D_2	D_3	D_4	Supply
S_1	10	8	4	3	500
S_2	12	14	20	2	400
S_3	6	9	25	25	300
Demand	250	350	600	150	

Table 4.20: Balanced the UCTP_1 by adding a dummy source

	D_1	D_2	D_3	D_4	Supply
S_1	10	8	4	3	500
S_2	12	14	20	2	400
S_3	6	9	25	25	300
D/S	0	0	0	0	150
Demand	250	350	600	150	

Step 2:

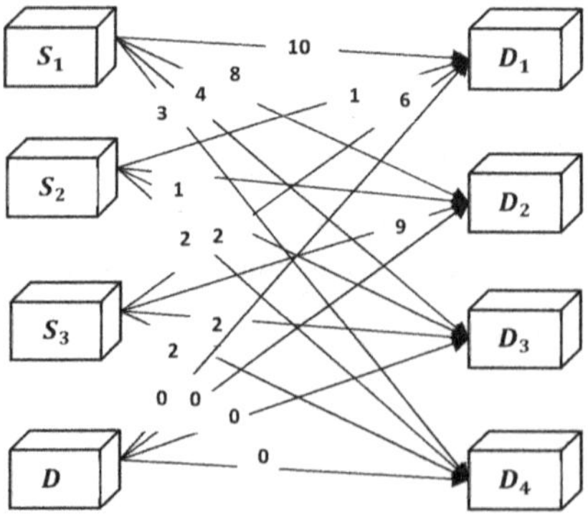

Figure 4.42: Transportation Network of UCTP_1

Step 3:

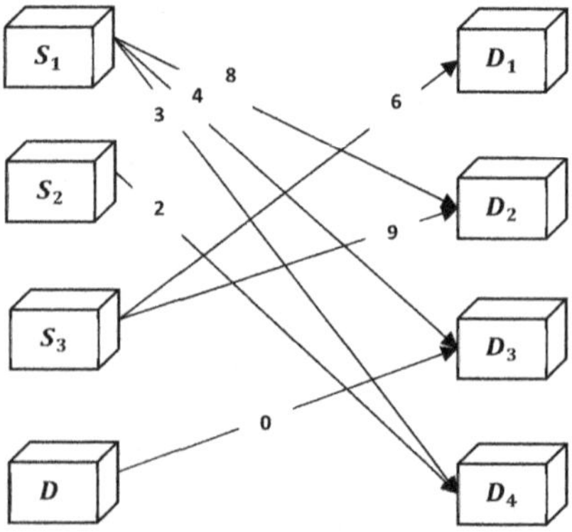

Figure 4.43: Minimum Spanning Tree of Transportation Network of UCTP_1

Step 4:

Allocation order: S_2D_4, S_1D_3, S_3D_1, S_3D_2

Steps 5 – 8:

Table 4.21: Allocation table of UCTP_1

	D_1	D_2	D_3	D_4	Supply
S_1	10	8	4 (500)	3	~~500~~,0
S_2	12	14 (250)	20	2 (150)	~~400~~,~~250~~,0
S_3	6 (250)	9 (50)	25	25	~~300~~,~~50~~,0
D/S	0	0 (50)	0 (100)	0	~~150~~,~~50~~,0
Demand	~~250~~,0	~~350~~,~~300~~,~~250~~,0	~~600~~,~~100~~,0	~~150~~,0	

Step 9:

Total transportation cost

$= (4 \times 500) + (14 \times 250) + (2 \times 150) + (6 \times 250) + (9 \times 50) = 7750$

The compression results of this example are in Section 5.3.2.

4.3.3. Example 3: Balanced Prohibited Route Transportation Problem_1 (BPRTP_1)[49]

Step 1:

Table 4.22: Initial Balanced Prohibited Route Transportation Problem (BPRTP_1)

	D_1	D_2	D_3	D_4	D_5	D_6	Supply
S_1	5	4	4	6	7	7	100
S_2	8	6	6	5	7	7	200
S_3	7	6	6	8	9	9	150
S_4	M	M	0	0	M	0	110
Demand	80	40	50	60	80	250	

Step 2:

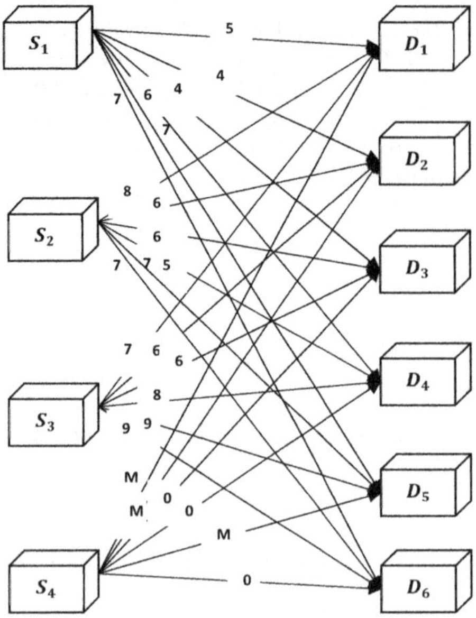

Figure 4.44: Transportation Network of BPRTP_1

Step 3:

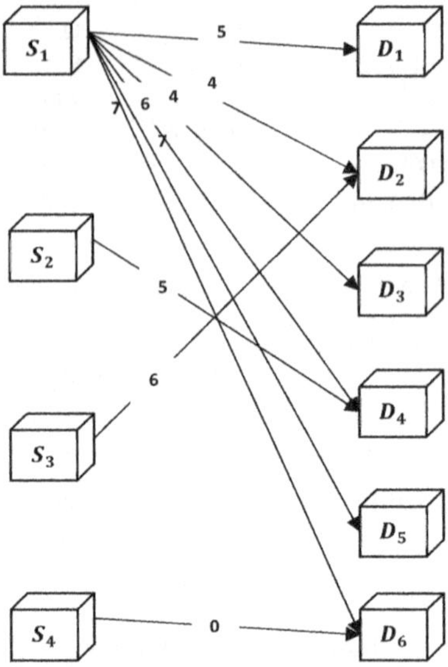

Figure 4.45: Minimum Spanning Tree of Transportation Network of BPRTP_1

Step 4:

Allocation order: S_1D_2, S_1D_3, S_1D_1, S_2D_4

Steps 5 – 8:

Table 4.23: Allocation table of BPRTP_1

	D_1	**D_2**	**D_3**	**D_4**	**D_5**	**D_6**	**Supply**
S_1	5 (10)	4 (40)	4 (50)	6	7	7	~~100~~, ~~60~~, ~~10~~, 0
S_2	8	6	6	5 (60)	7 (80)	7 (60)	~~200~~, ~~120~~, ~~60~~, 0
S_3	7 (70)	6	6	8	9	9 (80)	~~150~~, ~~80~~, 0
S_4	M	M	0	0	M	0 (110)	~~110~~, 0
Demand	~~80~~, ~~70~~, 0	~~40~~, 0	~~50~~, 0	~~60~~, 0	~~80~~, 0	~~250~~, ~~140~~, ~~80~~, 0	

Step 9:

Total transportation cost = $(5 \times 10) + (4 \times 40) + (4 \times 50) + (5 \times 60) + (7 \times 80) + (7 \times 60) + (7 \times 70) + (9 \times 80)$

$= 2900$

The compression results of this example are in Section 5.3.3.

4.3.4. Example 4: Unbalanced Prohibited Route Transportation Problem_1(UPRTP_1)[49]

Step 1:

Table 4.24: Initial Unbalanced Prohibited Route Transportation Problem (UPRTP_1)

	D_1	D_2	D_3	D_4	Supply
S_1	14	15	16	17	40
S_2	M	16	17	18	50
S_3	M	M	15	16	30
S_4	M	M	M	17	50
Demund	20	30	50	40	

Table 4.25: Balanced the UPRTP_1 by adding a dummy destination

	D_1	D_2	D_3	D_4	D/D	Supply
S_1	14	15	16	17	0	40
S_2	M	16	17	18	0	50
S_3	M	M	15	16	0	30
S_4	M	M	M	17	0	50
Demand	20	30	50	40	30	

Step 2:

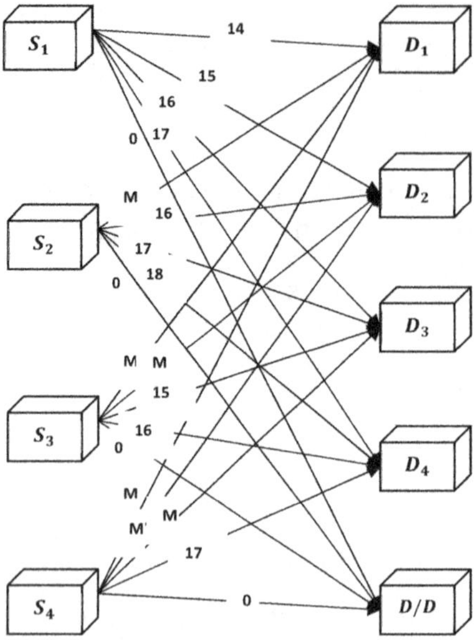

Figure 4.46: Transportation Network of UPRTP_1

Step 3:

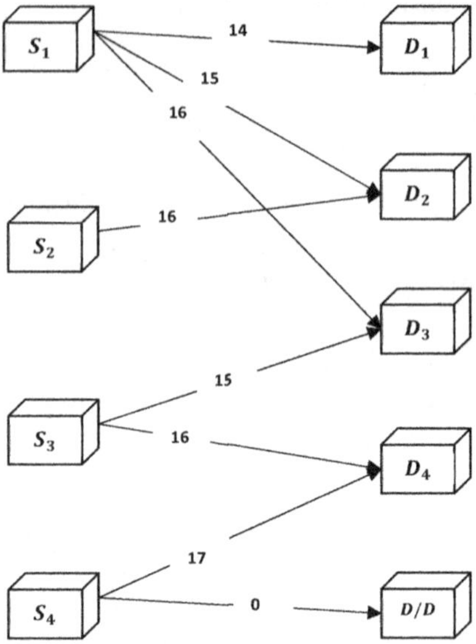

Figure 4.47: Minimum Spanning Tree of Transportation Network of UPRTP_1

Step 4: Allocation order: S_1D_1, S_3D_3, S_1D_2, S_2D_2, S_4D_4

Steps 5 – 8:

Table 4.26: Allocation table of UPRTP_1

	D_1	D_2	D_3	D_4	D/D	Supply
S_1	14 (20)	15 (20)	16	17	0	~~40~~,~~20~~,0
S_2	M	16 (10)	17 (20)	18	0 (20)	~~50~~,~~40~~,~~20~~,0
S_3	M	M	15 (30)	16	0	~~30~~,0
S_4	M	M	M	17 (40)	0 (10)	~~50~~,~~10~~,0
Demand	~~20~~,0	~~30~~,~~10~~,0	~~50~~,~~20~~,0	~~40~~,0	~~30~~,~~20~~,0	

Step 9:

Total transportation cost

$= (14 \times 20) + (15 \times 20) + (16 \times 10) + (17 \times 20) + (15 \times 30) + (17 \times 40)$

$= 2210$

The compression results of this example are in Section 5.3.4.

The two types of both balanced and unbalanced TPs are demonstrated in the examples above. CTP and PRTP, for instance. In every TP, the suggested approach provides an optimal solution, and when compared to the VAM method, it provides better compression.

CHAPTER 5

Discussion

This paper proposed a novel heuristic algorithm to solve MST problem using MACOA. Then applied this algorithm to IDAN and after that developed MACOA to solve both balanced and unbalanced classical and prohibited route TPs. In the result section, solve some numerical examples using proposed MST and TP algorithms. In this section, the obtained results of examples in the previous section are comparatively discussed. First, examples of MST are discussed. Then, a comparative analysis of the MST of IDAN with the proposed algorithm. Finally, the example results of TPs compare with other existing methods.

5.1 Comparative Analysis of the Proposed Modified Ant Colony Optimization Algorithm

The results obtained in the Modified Ant Colony Optimization (MACO) Algorithm of Minimum Spanning Tree examples 1 and 2 mentioned in 4.1.1 and 4.1.2 can compare with Prim's and Kruskal's algorithms.

Table 5.1: Comparative Analysis of example 1 and 2

Example No.	Prim's Algorithm	Kruskal's Algorithm	MACO Algorithm
1	84	84	84
2	132	132	132

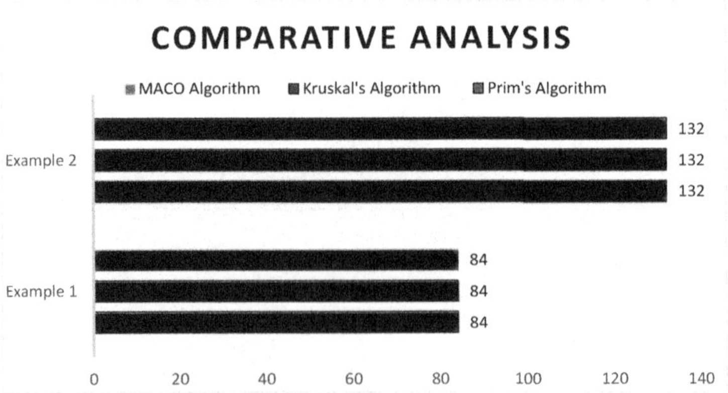

Figure 5.1: Comparative Analysis of Example 1 and 2 with Prim's, Kruskal's and MACO Algorithm

Table 5.1 and Figure 5.1 shows a comparative analysis of the proposed method. When compared to Prim's and Kruskal's algorithms same results can obtain. The total weight of example 1 minimum spanning tree is 84 units and example 2 is 132 units. An identical result may be produced using the algorithms of Prim and Kruskal.

5.2 Comparative Analysis of the Minimum Spanning Tree of Indian Domestic Air Network

In section 4.2, the proposed MACO algorithm is applied to find the MST of IDAN. There are 73 nodes and 140 edges in that network. After applying the proposed MACO algorithm the MST of that network can be obtained. The total weight of the Indian Domestic Air Network's Minimum Spanning Tree using the proposed method is equal to 29572.65 km. That result can be compared with other existing methods such as Prim's and Kruskal's algorithms.

Table 5.2. Comparative Analysis of the proposed method with Prim's and Kruskal's algorithms

Prim's Algorithm	Kruskal's Algorithm	MACO Algorithm
29572.65 km	29572.65 km	29572.65 km

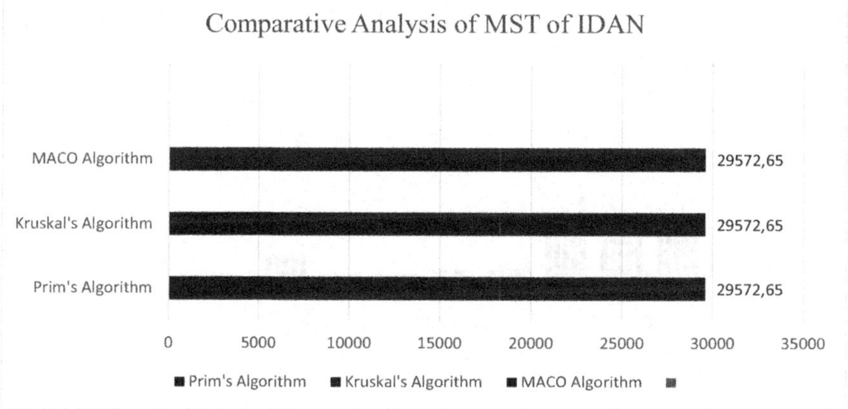

Figure 5.2. Comparative Analysis of the proposed method with Prim's and Kruskal's algorithms

Figure 5.2 and Table 5.2 above demonstrate the effectiveness of the new approach. Similar results may be obtained when compared to Prim's and Kruskal's methods. The Prim and Kruskal algorithms can generate an identical result.

5.3 The Comparative Analysis of the Proposed Transportation Problem Algorithm

5.3.1. Balanced Classical Transportation Problems (BCTPs)

This study examines the results' comparisons to assess the suggested approach. Table 5.3 shows the comparative analysis of Balanced Classical Transportation Problems (BCTP) with NWCM, LCM, VAM, and the proposed New Method (NEWM) for 10 instances. The problem solved in section 4.3.1 is comparatively analyzed in Table 5.3 and Figure 5.3.

Table 5.3: Comparative analysis of Balanced Classical Transportation Problems (BCTP)

Problem No. [46]	NWCM	LCM	VAM	NEWM	Optimal
BCTP_1	4160	3500	3320	3320	3320
BCTP_2	4400	2900	2850	2850	2850
BCTP_3	5925	4550	5125	4600	4525
BCTP_4	730	555	555	555	555
BCTP_5	520	475	475	475	435
BCTP_6	139	103	103	103	103
BCTP_7	226	156	156	156	156
BCTP_8	1500	1450	1500	1450	1390
BCTP_9	363	295	290	290	290
BCTP_10	234	191	187	187	183

Figure 5.3 shows the outcomes of the bar graphs used to display the comparison results found in Table 5.3 in more detail.

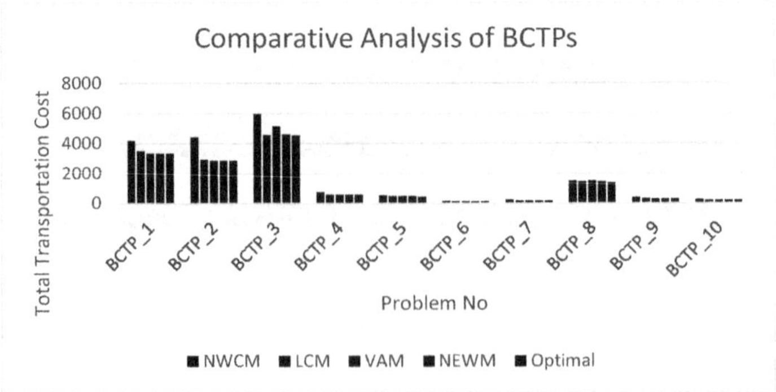

Figure 5.3: Comparative analysis of Balanced Classical Transportation Problems (BCTPs)

As can be seen from Table 5.3 above, the new approach is more effective than NWCM, LCM, and VAM in every situation where an increase in efficiency was achievable. Six out of ten instances give the optimal solution and other instances give the near-optimal solution. The numerical data of Table 5.3 is represented in Appendix C.

5.3.2. Unbalanced Classical Transportation Problems (UCTPs)

Table 5.4 shows the comparative analysis of Unbalanced Classical Transportation Problems (UCTP) with NWCM, LCM, VAM, and NEWM for 10 instances and Figure 5.4 shows its graphical representation. The problem solved in section 4.3.2 is comparatively analyzed in Table 5.4 and Figure 5.4.

Table 5.4: Comparative analysis of Unbalanced Classical Transportation Problems (UCTP)

Problem No. [33, 46, 47, 36]	**NWCM**	**LCM**	**VAM**	**NEWM**	**Optimal**
UCTP_1	18800	7750	8350	7750	7750
UCTP_2	1815	1695	1745	1695	1650
UCTP_3	12825	10375	11125	10375	10375
UCTP_4	13100	9200	9200	9200	9200
UCTP_5	1125	965	965	965	960
UCTP_6	8150	5750	6000	5750	5600
UCTP_7	5070	4900	5020	4900	4720
UCTP_8	1780	1465	1555	1465	1465
UCTP_9	1010	840	880	840	840
UCTP_10	788	606	630	606	606

Figure 5.4 shows the outcomes of the bar graphs used to display the comparison results found in Table 5.4 in more detail.

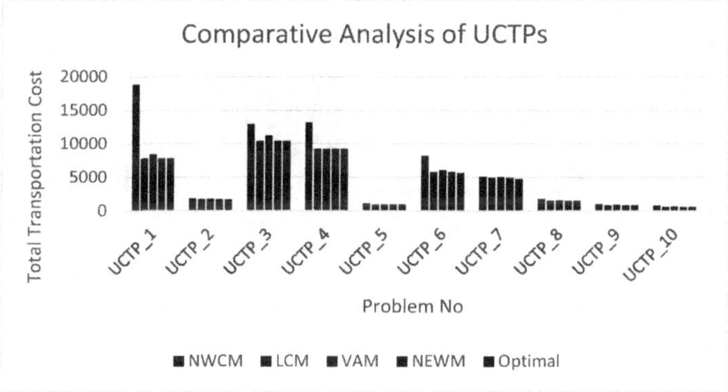

Figure 5.4: Comparative analysis of Unbalanced Classical Transportation Problems (UCTPs)

The new strategy outperforms NWCM, LCM, and VAM in every circumstance where a gain in efficiency was possible, as shown in Table 5.4 and Figure 5.4 above. The optimal answer is provided in six out of ten cases, and other cases provide a nearly optimal solution. The numerical data of Table 5.4 is represented in Appendix D.

5.3.3. Prohibited Route Transportation Problems (PRTPs)

Table 5.5 shows the comparative analysis of both Balanced Prohibited Route Transportation Problems (BPRTP) and Unbalanced Prohibited Route Transportation Problems (UPRTP) for five instances. The problems solved in sections 4.3.3 and 4.3.4 are comparative analyses in Table 5.5 and Figure 5.5.

Table 5.5: Comparative analysis of both balanced and unbalanced Prohibited Route Transportation Problems (PRTP)

Problem No. [49, 52]	**NWCM**	**LCM**	**VAM**	**NEWM**	**Optimal**
BPRTP_1	80012120	10260	2900	2900	2900
UPRTP_1	2230	2210	2230	2210	2210
UPRTP_2	39300	36100	36800	36800	36100
UPRTP_3	10400	9100	9100	9100	9100
UPRTP_4	11635	11565	11565	11565	11565

Figure 5.5 shows the outcomes of the bar graphs used to display the comparison results found in Table 5.5 in more detail. The numerical data of Table 5.5 is represented in Appendix E.

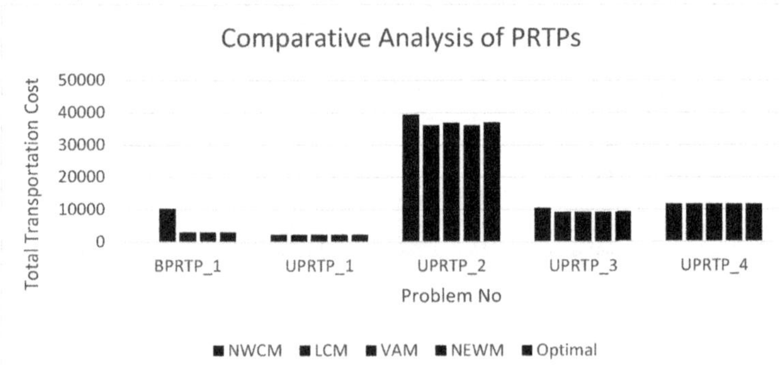

Figure 5.5: Comparative analysis of both balanced and unbalanced Prohibited Route Transportation Problems (PRTPs)

According to the data above (Table 5.5, Figure 5.5), the new approach is superior to the current methods. In four of the five cases, it delivers the best outcomes, while the other result is nearly as good as the VAM approach. Table 5.5 displays the comparison of randomly generated 10 instances of both CTPs and PRTPs.

5.3.4. Randomly Generated Transportation Problems

Table 5.6: Comparative analysis of randomly generated both classical and prohibited route transportation problems

Problem No.	NWCM	LCM	VAM	NEWM	Optimal
CTP_1	4125	3500	3500	3500	3500
CTP_2	9050	9275	9275	9275	8400
CTP_3	4150	3850	2900	2850	2850
CTP_4	15925	13150	12825	10650	10650
CTP_5	30700	14700	20400	14700	14700
PRTP_1	11665	11665	11665	11665	11665
PRTP_2	40550	31300	32150	31300	30650
PRTP_3	1215	1215	1215	1215	1215
PRTP_4	8050	6400	6400	6400	6400
PRTP_5	23400	23250	23250	23250	23250

Figure 5.6 shows the comparison data from Table 5.6 as well as the bar graphs that were used to display them.

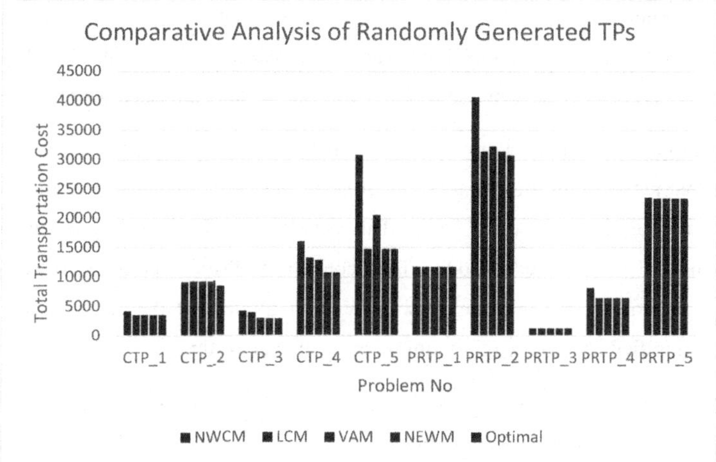

Figure 5.6: Comparative analysis of randomly generated both classical and prohibited route transportation problems

The suggested approach (NEWA) is eight out of ten times more effective than NWCM, LCM, and VAM, as is seen in Table 5.6 and Figure 5.6. The other two situations provide close to optimal answers, but those are better than VAM solutions. The numerical data of Table 5.6 is represented in Appendix F.

CHAPTER 6

Conclusion & Future Works

First, in this paper, a novel definite technique for determining the Minimum Spanning Tree (MST) of the weighted undirected graph G utilizing the ant colony algorithm on a given connected graph was developed. Starting with a single node, the Minimum Spanning Tree (MST) finds all of its accessible nodes as well as the set of connections that connect them with the least amount of weight.

A population-based metaheuristic called the Ant Colony Optimization (ACO) algorithm is useful for estimating solutions to complex problems. The proposed approach has been developed through the modification of the ACO algorithm and the addition of the Transition Rule. When the graph has a large number of nodes, this new algorithm is simpler to implement than other ones that already exist. Two illustrative examples were used to illustrate the newly suggested strategy. When compared to Prim's and Kruskal's methods, we can get the same results. The research also examined the problem at a large scale and introduced a novel easy and effective method to determine the MST of an undirected connected weight graph.

Then the proposed method was applied to the Indian Domestic Air Network (IDAN) to find the MST of that network. In this scenario, 73 airports and 140 airlines were examined. We take airports as nodes and airlines as edges. This network is a large-scale problem because the network has a large number of nodes. Then compare to Prim's and Kruskal's algorithm our proposed method gives the same answer. We can implement the results easily using the proposed modified ant colony minimum spanning tree algorithm.

At the end of this investigation, the proposed approach is more developed and provides a nearly optimum TP solution. Numerous strategies have been put out in the literature to address the transportation problem. Some approaches focused on finding an initial basic feasible solution, while others looked for the optimal possible solution. But utilizing Kruskal's algorithm, this study suggested a novel method for resolving transportation issues on both classical and prohibited routes. The minimum spanning tree of an undirected graph may be found using Kruskal's approach. Both balanced and unbalanced transportation issues may be solved using the suggested technique. The approach is also easy to understand and after a few repetitions, it offers us optimum or very close to optimal solutions.

In this study, 35 problems were addressed using this strategy. 10 out of 35 examples are generated at random, while the remaining 25 examples are selected from the literature. These

35 problems were examined in comparison to other techniques that were already in use. Show that, in comparison to the methods this research analyzed, the unique technique is effective in terms of the quality of the solution.

Mathematics plays a vital role in Science and Engineering. Researchers have proposed and bridged many areas in Mathematics with concepts in Fuzzy. In real-world problems, optimization techniques are useful for solving problems like project schedules, assignment problems, and network flow analysis. The main aspect of future work is to find the least transportation cost of some commodities through a capacitated network when the supply and demand of nodes and the capacity and cost of edges are represented as fuzzy numbers and crispy numbers. Also, plan the transportation problem using the ACO method and Genetic algorithm, and find the generalized solution. Moreover, several algorithms will be proposed to address the Assignment Problems and Travelling Salesman Problems.

REFERENCES

[1] J. Nešetřil, E. Milková, and H. Nešetřilová, "Otakar Borůvka on minimum spanning tree problem: Translation of both the 1926 papers, comments, history," *Discrete Math.*, vol. 233, no. 1–3, pp. 3–36, Apr. 2001, doi: 10.1016/S0012-365X(00)00224-7.

[2] U. Ekanayake, W. Daundasekara, and P. Perera, "Research An Approach for Solving Minimum Spanning Tree Problem and Transportation Problem Using Modified Ant Colony Algorithm," *North Am. Acad. Res.*, vol. 3, no. 9, pp. 28–45, 2020, doi: 10.5281/zenodo.4072472.

[3] M.-B. Choi and S.-U. Lee, "An Efficient Implementation of Kruskal's and Reverse-Delete Minimum Spanning Tree Algorithm," *J. Inst. Webcasting, Internet Telecommun.*, vol. 13, no. 2, pp. 103–114, Apr. 2013, doi: 10.7236/JIIBC.2013.13.2.103.

[4] H. N. Gabow, Z. Galil, T. Spencer, and R. E. Tarjan, "Efficient algorithms for finding minimum spanning trees in undirected and directed graphs," *Combinatorica*, vol. 6, no. 2, pp. 109–122, Jun. 1986, doi: 10.1007/BF02579168.

[5] S. Ruzika and H. W. Hamacher, "A survey on multiple objective minimum spanning tree problems," *Lect. Notes Comput. Sci. (including Subser. Lect. Notes Artif. Intell. Lect. Notes Bioinformatics)*, vol. 5515 LNCS, pp. 104–116, 2009, doi: 10.1007/978-3-642-02094-0_6/COVER.

[6] D. Vijayalakshmir and R. Kalaivani, "Minimum Cost Spanning Tree using Matrix Algorithm," *Int. J. Sci. Res. Publ.*, vol. 4, no. 9, 2014.

[7] P. Biswas, M. Goel, H. Negi, and M. Datta, "An Efficient Greedy Minimum Spanning Tree Algorithm Based on Vertex Associative Cycle Detection Method," *Procedia Comput. Sci.*, vol. 92, pp. 513–519, Jan. 2016, doi: 10.1016/J.PROCS.2016.07.376.

[8] M. Kirişci and D. Öncel, "The New Algorithm Involving Minimum Spanning Tree For Computer Networks In A Growing Company," *Istanbul Commer. Univ. J. Sci. İstanbul Ticaret Üniversitesi Fen Bilim. Derg.*, vol. 16, no. 32, pp. 1–14, 2017.

[9] S. I. Akbar, "Minimum Spanning Tree Algorithms And Techniques," *IJCIRAS*, vol. 1, no. 8, 2018.

[10] A. Khan, A. A. Aesha, and J. Sarker, "A new algorithmic approach to finding a minimum spanning tree," *4th Int. Conf. Electr. Eng. Inf. Commun. Technol. iCEEiCT 2018*, pp. 590–594, Jan. 2019, doi: 10.1109/CEEICT.2018.8628095.

[11] A. Dey, S. Broumi, L. H. Son, A. Bakali, M. Talea, and F. Smarandache, "A new algorithm for finding minimum spanning trees with undirected neutrosophic graphs," *Granul. Comput.*, vol. 4, no. 1, pp. 63–69, Jan. 2019, doi: 10.1007/S41066-018-0084-7.

[12] P. S. Reddy and A. M. Derdar, "Minimal Spanning Tree," *International Journal of Engineering Research & Technology (IJERT),* vol. 6, no. 3, pp. 164–168, 2017.

[13] S. Pettie and V. Ramachandran, "An Optimal Minimum Spanning Tree Algorithm," *Journal of the ACM*, vol. 49, no. 1, pp. 16–34, 2022.

[14] S. Majumder, P. S. Barma, A. Biswas, P. Banerjee, B. K. Mandal, S. Kar, and P. Ziemba, "On Multi-Objective Minimum Spanning Tree Problem under Uncertain Paradigm," *Symmetry* 2022, 14, 106. https://doi.org/ 10.3390/sym14010106

[15] Paryati and K. Salahddine, “The Implementation of Kruskal’s Algorithm for Minimum Spanning Tree in a Graph,” *MATEC Web Conf.*, vol. 348, p. 01001, 2021, doi: 10.1051/matecconf/202134801001.

[16] A. SAYLI and J. H. S. Alkhalissi, “Negligence Minimum Spanning Tree Algorithm,” *Eur. J. Sci. Technol.*, no. November, pp. 70–76, 2018, doi: 10.31590/ejosat.386716.

[17] R. Abhilasha, “Minimum Cost Spanning Tree Using Prim’s Algorithm,” *International Journal of Advanced Research in Computer Science and Management Studies*, vol. 1, no. 1, pp. 15–20, 2013.

[18] P. Ayegba, J. Ayoola, E. Asani, and A. Okeyinka, “A Comparative Study of Minimal Spanning Tree Algorithms,” *2020 Int. Conf. Math. Comput. Eng. Comput. Sci. ICMCECS 2020*, no. May 2020, doi: 10.1109/ICMCECS47690.2020.240900.

[19] M. Dorigo, V. Maniezzo, and A. Colorni, “Dorigo-Maniezzo Colomi_the-Ant-System-Optimization-By-a-Colony-of-Cooperating-Agents,” *IEEE Trans. Syst. Man, Cybern. B*, vol. 26, no. 1, pp. 1–26, 1999.

[20] E. M. U. S. B. Ekanayake, "An Improved Ant Colony Algorithm to Solve Prohibited Transportation Problems," *International Journal of Applied Mathematics and Theoretical Physics*, vol. 8, no. 2, 2022, pp. 43-51. doi: 10.11648/j.ijamtp.20220802.12

[21] F. L. Hitchcock, “The Distribution of a Product from Several Sources to Numerous Localities,” *J. Math. Phys.*, vol. 20, no. 1–4, pp. 224–230, Apr. 1941, doi: 10.1002/SAPM1941201224.

[22] T. C. Koopmans, “Optimum Utilization of the Transportation System," *Econometric Society* vol. 17, pp. 136–146, 1941.

[23] H. A. Taha, “Operations research : an introduction,” *Pearson/Prentice Hall*, p. 813, 2007.

[24] Z. A. M. S. Juman and N. G. S. A. Nawarathne, “An efficient alternative approach to solving a transportation problem,” *Ceylon J. Sci.*, vol. 48, no. 1, p. 19, 2019, doi: 10.4038/cjs.v48i1.7584.

[25] S. Korukoğlu and S. Ballı, “An Improved Vogel’s Approximation Method For The Transportation Problem,” *Math. Comput. Appl.*, vol. 16, no. 2, pp. 370–381, 2011.

[26] A. Quddoos, S. Javaid, M.M. Khalid, "A New Method To Solve Bi-Objective Transportation Problem," *International Journal of Applied Mathematics,* vol. 26, no. 5, pp. 555–563, 2013.

[27] P. Adhikari and G. B. Thapa, “A Note on Feasibility and Optimality of Transportation Problem,” *J. Inst. Eng.*, vol. 10, no. 1, pp. 59–68, 2014, doi: 10.3126/jie.v10i1.10879.

[28] A. Edward Samuel, “Improved zero point method (IZPM) for transportation problems,” *Appl. Math. Sci.*, vol. 6, no. 109–112, pp. 5421–5426, 2012.

[29] D. A. S. Kadhem and M. S. Taher, “Maximum { Supplies, Demands } Method To Find The Initial Transportation Problem,” *Journal of University of Zakho*, vol. 1, no. 2, pp. 849–853, 2013.

[30] A. Arockiamary and G.Pravina, "Application of Graph Theory to Find Shortest Path of Transportation Problem," *International Journal of Computing Algorithm,* vol. 3, pp. 833–837, 2014.

[31] M. M. Ahmed, A. R. Khan, F. Ahmed, and M. S. Uddin, “Incessant Allocation Method

for Solving Transportation Problems," *Am. J. Oper. Res.*, vol. 06, no. 03, pp. 236–244, 2016, doi: 10.4236/AJOR.2016.63024.

[32] S. V. Kamble and B. G. Kore, "A New Method to Obtain an Initial Basic Feasible Solution of Transportation Problem with the Average Opportunity Cost Method," *Int. J. Eng. Adv. Technol.*, vol. 9, no. 2, pp. 206–209, 2019, doi: 10.35940/ijeat.f9283.129219.

[33] A. K. M Selim Reza, A. R. M Jalal Uddin Jamali, and B. Biswas, "A Modified Algorithm for Solving Unbalanced Transportation Problems," *J. Eng. Sci.*, vol. 10, no. 1, pp. 93–101, 2019.

[34] A. Babu, A. Helal, M. S. Hasan, and U. K. Das, "Implied Cost Method (ICM): An Alternative Approach to Find the Feasible Solution of Transportation Problem," *Type Double Blind Peer Rev. Int. Res. J. Publ. Glob. Journals Inc*, vol. 14, no. 1, pp. 5–13, 2014.

[35] S. Sood and K. Jain, "The Maximum Difference Method to Find Initial Basic Feasible Solution for Transportation Problem," *Asian J. Manag. Sci.*, vol. 03, no. 07, pp. 8–11, 2015.

[36] T. Geetha and N. Anandhi, "Method for Solving Unbalanced Transportation Problems Using Standard Deviations," *Int. J. Pure Appl. Math.*, vol. 119, no. 16, pp. 4971–4989, 2018.

[37] R.G. Patel, D. P. H. Bhathawala, and D.B.S. Patel, "'An Alternate Approach to Find an Optimal Solution of a Transportation Problem.,'" *IOSR J. Math.*, vol. 13, no. 01, pp. 01–05, 2017, doi: 10.9790/5728-1301030105.

[38] O. Jude, O. Ben Ifeanyichukwu, I. A. Ihuoma, and E. P. Akpos, "A New and Efficient Proposed Approach to Find Initial Basic Feasible Solution of a Transportation Problem," *Am. J. Appl. Math. Stat.*, vol. 5, no. 2, pp. 54–61, 2017, doi: 10.12691/ajams-5-2-3.

[39] S. Khan, Aminur & Adrian, Vilcu & Sultana, Nahid & Ahmed, "Determination of Initial Basic Feasible Solution of a Transportation Problem: A TOCM-SUM Approach," *Bul. Institutului Politeh. Din Iasi, Rom. Secţia Autom. si Calc. . .*, vol. LXI (LXV), no. Lxv, pp. 39–49, 2015.

[40] M. P. Rakshit, "Another approach for obtaining the initial feasible solution of Transportation Problem," *Res. Rev. Int. J. Multidiscip.*, vol. 4, no. 5, 2019.

[41] S. Uddin and A. R. Khan, "Improved Least Cost Method to Obtain a Better IBFS to the Transportation Problem," vol. 6, no. 2, pp. 1–20, 2016.

[42] K. Venkatasubbaiah, S. G. Acharyulu, and C. Mouli, "Fuzzy Goal Programming Method for SolvingMulti-Objective Transportation Problems Fuzzy Goal Programming Method for Solving Multi-Objective Transportation Problems," *Type Double Blind Peer Rev. Int. Res. J. Publ. Glob. Journals Inc*, vol. 11.

[43] Z. A. M. Silmi Juman, M. Masoud, M. Elhenawy, H. Bhuiyan, M. M. R. Komol, and O. Battaïa, "A new algorithm for solving uncapacitated transportation problem with interval-defined demands and suppliers capacities," *J. Intell. Fuzzy Syst.*, vol. 41, no. 1, pp. 625–637, 2021, doi: 10.3233/JIFS-202436.

[44] Z. A. M. S. Juman *et al.*, "Close Interval Approximation of Pentagonal Fuzzy Numbers for Interval Data-Based Transportation Problems," *Sustain.*, vol. 14, no. 12, 2022, doi:

10.3390/su14127423.

[45] E. M. U. S. B. Ekanayake, S. P. C. Perera, W. B. Daundasekara, and Z. A. M. S. Juman, "An Effective Alternative New Approach in Solving Transportation Problems," *Am. J. Electr. Comput. Eng.*, vol. 5, no. 1, p. 1, 2021, doi: 10.11648/j.ajece.20210501.11.

[46] E. M. U. S. B. Ekanayake, W. B. Daundasekara, and S. P. C. Perera, "Solution of a Transportation Problem using Bipartite Graph," *Glob. Journals*, vol. 21, no. October 2021.

[47] E. M. U. S. B. Ekanayake, S. P. C. Perera, W. B. Daundasekara, and Z. A. M. S. Juman, "A Modified Ant Colony Optimization Algorithm for Solving a Transportation Problem," *J. Adv. Math. Comput. Sci.*, no. August, pp. 83–101, 2020, doi: 10.9734/jamcs/2020/v35i530284.

[48] E. M. U. S. B. Ekanayake, W. B. Daundasekara, and S. P. C. Perera, "New Approach to Obtain the Maximum Flow in a Network and Optimal Solution for the Transportation Problems," *Mod. Appl. Sci.*, vol. 16, no. 1, p. 30, 2022, doi: 10.5539/mas.v16n1p30.

[49] E. M. U. S. B. Ekanayake, "An Improved Ant Colony Algorithm to Solve Prohibited TransportationProblems," *International Journal of Applied Mathematics and Theoretical Physics*, vol. 8, no. 2, 2022, pp. 43-51.doi: 10.11648/j.ijamtp.20220802.12

[50] E. M. U. S. B. Ekanayake, W. B. Daundasekara, S. P. C. Perera, "An Examination of Different Types of Transportation Problems and Mathematical Models," *American Journal of Mathematical and Computer Modelling*, vol. 7, no. 3, 2022, pp. 37-48. doi: 10.11648/j.ajmcm.20220703.11

[51] E. M. U. S. B. Ekanayake, "A Novel Approach Algorithm for Determining the Initial Basic Feasible Solution for Transportation Problems," pp. 234–246, 2022, doi: 10.47540/ijias.v2i3.529.

[52] E. M. D. B. Ekanayake and E. M. U. S. B. Ekanayake, "Performance Of The Best Solution For The Prohibited Route Transportation Problem By An Improved Vogel's Approximation Method," *Indonesian Journal of Applied Research (IJAR)*, vol. 3, no. 3, pp. 190–206, 2022, doi: 10.30997/ijar.v3i3.241.

[53] J. Ackora-Prah, V. Acheson, B. Barnes, I. Takyi, and E. Owusu-Ansah, "A Two-Phase Method for Solving Transportation Models with Prohibited Routes," *Pakistan J. Stat. Oper. Res.*, vol. 18, no. 3, pp. 749–758, 2022, doi: 10.18187/pjsor.v18i3.3911.

[54] W. K. Abdelbasset and S. H. Elsayed, "Study on cross-border fresh order and transport model based on profit maximization principle," *Food Sci. Technol.*, vol. 42, pp. 1–7, 2022, doi: 10.1590/fst.75821.

[55] A. Seethalakshmy and N. Srinivasan, "Solving a real-life problem using transportation," *Int. J. Innov. Technol. Explor. Eng.*, vol. 8, no. 10, pp. 4284–4287, 2019, doi: 10.35940/ijitee.J1062.0881019.

[56] S. K. Prabha, "New Algorithm via Method of Magnitude To Maximize the Profit of a Fuzzy Transportation Problem," *ISROJ Publication*, vol. 01, no. 05, pp. 1–6, 2016.

[57] K. P. O. Niluminda and E. M. U. S. B. Ekanayake, "Innovative Matrix Algorithm to Address the Minimal Cost-Spanning Tree Problem," *J. Electrical Electron Eng.*, vol. 1, no. 1, pp. 148–153, 2022.

[58] K. P. O. Niluminda and E. M. U. S. B. Ekanayake, "Kruskal's algorithm for solving the both balanced unbalanced acceptable and prohibited route transportation problems," *North American Academic Research*, vol. 5, no. 12, 2022.

[59] K. P. O. Niluminda and E. M. U. S. B. Ekanayake, "Novel approach based on the weight adjacency matrix to obtain the minimum weight spanning tree," *Proceedings of FARS2022*, vol. 1, 2022.

[60] K. P. O. Niluminda and E. M. U. S. B. Ekanayake, "An efficient method to solve minimum spanning tree problem using graph theory and improved ant colony optimization algorithm," *North American Academic Research*, vol. 5, no. 12, 2022.

[61] S. Krishna Prabha and S. Vimala, "A modified method for solving a balanced fuzzy transportation problem for maximizing the profit," *Int. J. Pure Appl. Math.*, vol. 106, no. 5, pp. 45–52, 2016, doi: 10.12732/ijpam.v106i5.5.

[62] S. Jana, B. Das, G. Panigrahi, and M. Maiti, "Profit Maximization Solid Transportation Problem with Gaussian Type-2 Fuzzy Environments," *Ann. Pure Appl. Math.*, vol. 16, no. 2, pp. 323–335, 2018, doi: 10.22457/apam.v16n2a9.

[63] H. J. Kadhim, A. K. Mushtak Shiker, and H. A. H. Al-Dallal, "New Technique for Finding the Maximization to Transportation Problems," *J. Phys. Conf. Ser.*, vol. 1963, no. 1, 2021, doi: 10.1088/1742-6596/1963/1/012070.

[64] A. Kalam Azad, "An Innovative Algorithmic Approach for Solving Profit Maximization Problems," *Math. Lett.*, vol. 4, no. 1, p. 1, 2018, doi: 10.11648/j.ml.20180401.11.

[65] M. S. F. Kader and H. Alam, "Use of a New Transportation Algorithm for Profit Maximization," *Eur. Sci. J. ESJ*, vol. 15, no. 15, pp. 262–267, 2019, doi: 10.19044/esj.2019.v15n15p262.

[66] D. S. Kumar and G. C. Rabinson, "Profit Maximization of Balanced Fuzzy Transportation Problem Using Ranking Method," *J. Comput. Math. Sci.*, vol. 9, no. 7, pp. 722–726, 2018, doi: 10.29055/jcms/810.

[67] B. Sinha, A. Das, and U. K. Bera, "Profit Maximization Solid Transportation Problem with Trapezoidal Interval Type-2 Fuzzy Numbers," *Int. J. Appl. Comput. Math.*, vol. 2, no. 1, pp. 41–56, 2016, doi: 10.1007/s40819-015-0044-8.

[68] P. K. Giri, M. K. Maiti, and M. Maiti, "Profit maximization solid transportation problem under budget constraint using fuzzy measures," *Iran. J. Fuzzy Syst.*, vol. 13, no. 5, pp. 35–63, 2016.

[69] M. A. Nomani, I. Ali, and A. Ahmed, "A new approach for solving multi-objective transportation problems," *Int. J. Manag. Sci. Eng. Manag.*, vol. 12, no. 3, pp. 165–173, 2017, doi: 10.1080/17509653.2016.1172994.

[70] K. Singh, S. Rajan, "Geometric Mean Method to Solve Multi-Objective Transportation Problem Under Fuzzy Environment," *International Journal of Innovative Technology and Exploring Engineering (IJITEE)*, vol. 9, no. 5, pp. 1739-1744, 2020, doi: 10.35940/ijitee.E2859.039520.

[71] M. H. Lohgaonkar and V. H. Bajaj, "A fuzzy approach to solve multi-objective transportation problem," *Int. J. Agric. Stat. Sci.*, vol. 5, no. 2, pp. 443–452, 2009.

[72] K. Bharathi and C. Vijayalakshmi, "Optimization of multi-objective transportation problem using evolutionary algorithms," *Glob. J. Pure Appl. Math.*, vol. 12, no. 2, pp.

1387–1396, 2016.

[73] M. M. Ahmed, A. R. Khan, M. S. Uddin, and F. Ahmed, "A New Approach to Solve Transportation Problems," *Open J. Optim.*, vol. 05, no. 01, pp. 22–30, 2016, doi: 10.4236/ojop.2016.51003.

[74] M. Zangiabadi and H. R. Maleki, "Fuzzy goal programming technique to solve multiobjective transportation problems with some non-linear membership functions," *Iran. J. Fuzzy Syst.*, vol. 10, no. 1, pp. 61–74, 2013.

[75] D. M. Doke, "A Solution to Three Objective Transportation Problems Using Fuzzy Compromise Programming Approach," *International Journal of Modern Sciences and Engineering Technology (IJMSET)*, vol. 2, no. 9, pp. 9–13, 2015.

[76] S. R. Ahir and H. M. Tandel, "Solution of Multi Objective Transportation Problem," *International Journal of Trend in Scientific Research and Development Trend in Scientific Research and Development (IJTSRD)*, vol. 5, no. 4, pp. 1331–1337, 2021.

[77] P. Goel, "New Row Maxima Method To Solve Multi-Objective Transportation Problem Using C-Programme And Fuzzy Technique," *Int. J. Eng. Sci. Math.*, vol. 10, 2021.

[78] E. M. U. S. B. Ekanayake, "Geometric Mean Method Combined With Ant Colony Optimization Algorithm to Solve Multi-Objective Transportation Problems in Fuzzy Environments," *J. Electrical Electron Eng.*, vol. 1, no. 1, pp. 39–47, 2022.

[79] K. Singh and D. S. Rajan, "Geometric Mean Method to Solve Multi Objective Transportation Problem Under Fuzzy Environment," *Int. J. Innov. Technol. Explor. Eng.*, vol. 9, no. 5, pp. 1739–1744, 2020, doi: 10.35940/ijitee.e2859.039520.

[80] M. A. E. Afwat, A. A. M. Salama, and N. Farouk, "A New Efficient Approach to Solve Multi-Objective Transportation Problem in the Fuzzy Environment (Product approach)," *International Journal of Applied Engineering Research*, vol. 13, no. 18, pp. 13660–13664, 2018.

[81] Y. Kacher and P. Singh, "A Comprehensive Literature Review on Transportation Problems," *Int. J. Appl. Comput. Math*, vol. 7, no. 5, 2021. doi: 10.1007/s40819-021-01134-y.

[82] K. Singh and S. Rajan, "Matrix maxima method to solve multi-objective transportation problem with a pareto optimality criteria," *Int. J. Innov. Technol. Explor. Eng.*, vol. 8, no. 11, pp. 1929–1932, 2019, doi: 10.35940/ijitee.K2134.0981119.

[83] P. Pandian and D. Anuradha, "A new method for solving bi-objective transportation problems," *Aust. J. Basic Appl. Sci.*, vol. 5, no. 10, pp. 67–74, 2011.

[84] T. Karthy and K. Ganesan, "Multi Objective Transportation Problem - Genetic Algorithm Approach," *International Journal of Pure and Applied Mathematics*, vol. 119, no. 9, pp. 343–350, 2018.

[85] M. Afwat, A. A. M. Salama, and N. Farouk, "A New Efficient Approach to Solve Multi-Objective Transportation Problem in the Fuzzy Environment (Product approach)," *Int. J. Appl. Eng. Res.*, vol. 13, pp. 13660–13664, 2018.

[86] M. Zangiabadi and H. R. Maleki, "Fuzzy Goal Programming for Multiobjective," *Computing*, vol. 24, no. 1, pp. 449–460, 2007.

[87] M. Ogbi, "A Mathematical Property of the Harmonic Mean," *6th Int. Days Stat. Econ.*,

pp. 873–877, 2012.

[88] K. K. Jain, R. Bhardwaj, and S. Choudhary, "A multi-objective transportation problem solve by laxicographic goal programming," *Int. J. Recent Technol. Eng.*, vol. 7, no. 6, pp. 1842–1846, 2019.

[89] O. George A., "Solution of Multi-Objective Transportation Problem Via Fuzzy Programming Algorithm," *Sci. J. Appl. Math. Stat.*, vol. 2, no. 4, p. 71, 2014, doi: 10.11648/j.sjams.20140204.11.

[90] M. A. M. Khan and S. J. Kabeer, "Multi-Objective Transportation Problem Under Fuzziness with S-type Membership Function," no. 4, pp. 66–69, 2015.

[91] R. Likaj, A. Shala, M. Mehmetaj, P. Hyseni, and X. Bajrami, "Application of graph theory to find optimal paths for the transportation problem," *IFAC Proc. Vol.*, vol. 15, no. PART 1, pp. 235–240, 2013, doi: 10.3182/20130606-3-XK-4037.00031.

[92] PriyadharshiniT and ManikandanKM, "Application Of Graph Theory To Find Optimal Path And Minimized Cost For The Transportation Problem," *Int. J. Math. Trends Technol.*, vol. 65, 2019.

APPENDICES

Appendix A: Indian Domestic Airports with codes

No.	Airports	Codes
1.	Agartala	IXA
2.	Agatti	AGX
3.	Agra	AGR
4.	Ahmedabad	AMD
5.	Aizawl	AJL
6.	Allahabad	IXD
7.	Amritsar	ATQ
8.	Aurangabad	IXU
9.	Bagdogra	IXB
10.	Bengaluru	BLR
11.	Bhatinda	BUP
12.	Bhavnagar	BHU
13.	Bhopal	BHO
14.	Bhubaneswar	BBI
15.	Bhuj	BHJ
16.	Chandigarh	IXC
17.	Chennai	MAA
18.	Coimbatore	CJB
19.	Dehradun	DED
20.	Delhi	DEL
21.	Dharamshala	DHM
22.	Dibrugarh	DIB
23.	Dimapur	DMU
24.	Diu	DIU
25.	Gaya	GAY
26.	Goa	GOI
27.	Gorakhpur	GOP
28.	Guwahati	GAU
29.	Gwalior	GWL
30.	Hubli	HBX
31.	Hyderabad	HYD
32.	Imphal	IMF
33.	Indore	IDR
34.	Jabalpur	JLR
35.	Jaipur	JAI
36.	Jammu	IXJ
37.	Jamnagar	JGA

No.	Airports	Codes
38.	Jodhpur	JDH
39.	Kanpur	KNU
40.	Khajuraho	HJR
41.	Kochi	COK
42.	Kolkata	CCU
43.	Kozhikode	CCJ
44.	Kullu	KUU
45.	Leh	IXL
46.	Lilabari	IXI
47.	Lucknow	LKO
48.	Madurai	IXM
49.	Mangalore	IXE
50.	Mumbai	BOM
51.	Nanded	NDC
52.	Naqpur	NAG
53.	Pantnagar	PGH
54.	Pasighat	IXT
55.	Patna	PAT
56.	Port Blair	IXZ
57.	Pune	PNQ
58.	Raipur	RPR
59.	Rajkot	RAJ
60.	Ranchi	IXR
61.	Shillong	SHL
62.	Silchar	IXS
63.	Srinagar	SXR
64.	Surat	STV
65.	Tezpur	TEZ
66.	Tiruchirapally	TRZ
67.	Thiruvananthapuram	TRV
68.	Tirupati	TIR
69.	Udaipur	UDR
70.	Vadodara	BDQ
71.	Varanasi	VNS
72.	Vijayawada	VGA
73.	Visakhapatnam	VTZ

Appendix B: Name of the Air Routes with distances

Air Routes	Air Distance(Km)	Edges
CCJ – CJB	119.32	43 - 18
COK – CCJ	119.88	41 - 43
BOM – PNQ	124.19	50 - 57
IXC – KUU	138.23	16 - 44
AJL – IMF	164.91	5 - 32
DEL – AGR	178.08	20 - 3
COK – TRV	194.18	41 - 67
IXS – TEZ	200.66	62 - 65
DEL – DED	208.48	20 - 19
DMU – DIB	216.72	23 - 22
VNS – GAY	224.87	71 - 25
BOM – STV	225.59	50 - 64
DEL – JAI	231.69	20 - 35
DEL – IXC	236.24	20 - 16
DEL – PGH	236.74	20 - 53
VNS – LKO	244.63	71 - 47
IXL – SXR	255.89	45 - 63
HYD – VGA	263.62	31 - 72
MAA – BLR	267.74	17 - 10
BUP – IXJ	269.08	11 - 36
BOM – DIU	272.11	50 - 24
GAU – IMF	276.11	28 - 32
BHO – JLR	277.62	13 - 34
BOM – IXU	278.84	50 - 8
RPR – NAG	279.31	58 - 52
GAU – IXI	282.37	28 - 46
BLR – TRZ	291.65	10 - 66
MAA – TRZ	293.94	17 - 66
DEL – BUP	295.9	20 - 11
IXJ – IXL	298.48	36 - 45
VNS – HJR	304.31	71 - 40
BOM – BHU	304.58	50 - 12
BLR – IXE	306.14	10 - 49
TRV – CCJ	313.53	67 - 43
CCU – IXA	316.54	42 - 1
HJR – AGR	325.41	40 - 3
GAU – IXB	330.66	28 - 9
BLR – COK	367.24	10 - 41
DEL – KUU	368.09	20 - 44
BLR – HBX	371.03	10 - 30
VTZ – BBI	391.2	73 - 14
DEL – KNU	399.11	20 - 39

DEL – DHM	408.15	20 - 21
DEL – ATQ	413.85	20 - 7
MAA – IXM	417.47	17 - 48
HYD – TIR	417.56	31 - 68
BOM – RAJ	418.8	50 - 59
DEL – LKO	423.45	20 - 47
BOM – GOI	424.78	50 - 26
CCU – GAY	424.98	42 - 25
GAU – IXT	430.41	28 - 54
HYD – NAG	434.16	31 - 52
BOM – AMD	444.14	50 - 4
CCU – AJL	446.22	42 - 5
CCU – IXB	447.86	42 - 9
HYD – BLR	455.14	31 - 10
COK – AGX	468.02	41 - 2
DEL – JDH	475.79	20 - 38
BOM – HBX	476.59	50 - 30
BOM – JGA	478.59	50 - 37
BLR – GOI	482.91	10 - 26
CCU – SHL	493.18	42 - 61
BLR – VGA	497.57	10 - 72
CCU – GAU	498.29	42 - 28
HYD – PNQ	500.24	31 - 57
MAA – CCJ	502.38	17 - 43
DEL – IXJ	507.02	20 - 36
MAA – HYD	507.19	17 - 31
HYD – VTZ	511.4	31 - 73
MAA – COK	517.64	17 - 41
CCJ – GOI	524.87	43 - 26
CCU – IXS	524.99	42 - 62
DEL – UDR	542.57	20 - 69
BOM – BHJ	572.55	50 - 15
DEL – IXD	575.49	20 - 6
DEL – BHO	587.44	20 - 13
MAA – TRV	614.13	17 - 67
MAA – VTZ	619.65	17 - 73
DEL – IXL	620.69	20 - 45
BOM – UDR	623.85	50 - 69
CCU – TEZ	628.62	42 - 65
DEL – SXR	642	20 - 63
CCU – DMU	648.1	42 - 23
DEL – GOP	657.13	20 - 27
BOM – BHO	657.66	50 - 13
DEL – IDR	662.83	20 - 33

DEL – VNS	666.97	20 - 71
DEL – JLR	667.78	20 - 34
HYD – JLR	682.48	31 - 34
BOM – NAG	682.56	50 - 52
BOM – IXE	714.72	50 - 49
BLR – PNQ	722.65	10 - 57
MAA – GOI	733.04	17 - 26
DEL – AMD	756.48	20 - 4
CCU – IXI	768.33	42 - 46
DEL – BDQ	794.44	20 - 70
BOM – BLR	834.12	50 - 10
CCU – DIB	851.87	42 - 22
DEL – NAG	853.85	20 - 52
DEL – PAT	856.67	20 - 55
PNQ – RPR	866.97	57 - 58
DEL – GAY	887.83	20 - 25
BOM – JAI	911.28	50 - 35
CCU – IXT	916.74	42 - 54
DEL – STV	937.99	20 - 64
DEL – RAJ	941.53	20 - 59
BOM – CCJ	944.13	50 - 43
DEL – RPR	944.79	20 - 58
BOM – RPR	954.65	50 - 58
BOM – GWL	971.29	50 - 29
DEL – IXU	983.05	20 - 8
BOM – CJB	1001.74	50 - 18
MAA – BBI	1006.07	17 - 14
DEL – IXR	1007.62	20 - 60
BOM – COK	1063.76	50 - 41
BOM – VTZ	1102.99	50 - 73
DEL – IXB	1125.11	20 - 9
DEL – BOM	1138.18	20 - 50
BOM – LKO	1183.53	50 - 47
CCU – HYD	1207.61	42 - 31
VTZ – IXZ	1225.12	73 - 56
BOM – VNS	1247.23	50 - 71
BOM – TRV	1257.65	50 - 67
DEL – HYD	1267.62	20 - 31
DEL – BBI	1277.65	20 - 14
CCU – IXZ	1306.08	42 - 56
DEL – CCU	1312.62	20 - 42
DEL – VTZ	1357.93	20 - 73
BOM – BBI	1361.62	50 - 14
MAA – IXZ	1372.5	17 - 56

CCU – MAA	1385.64	42 - 17
ATQ – NDC	1415.43	51 - 7
DEL – GAU	1455.44	20 - 28
CCU – BLR	1546.77	42 - 10
BOM – CCU	1665.41	50 - 42
DEL – BLR	1709.97	20 - 10
DEL – MAA	1760.78	20 - 17
BLR – GAU	2039.29	10 - 28
DEL – COK	2048.81	20 - 41

Appendix C: Data set of balanced classical TP examples

Problem No. [46]	Data
BCTP_1	$c_{ij} = \{50, 60, 100, 50;\ 80, 40, 70, 50; 90, 70, 30, 50\}$ $S_{ij} = \{20, 38, 16\}$, $D_{ij} = \{10, 18, 22, 24\}$
BCTP_2	$c_{ij} = \{3, 1, 7, 4;\ 2, 6, 5, 9; 8, 3, 3, 2\}$ $S_{ij} = \{300, 400, 500\}$, $D_{ij} = \{250, 350, 400, 200\}$
BCTP_3	$c_{ij} = \{6, 8, 10;\ 7, 11, 11; 4, 5, 12\}$ $S_{ij} = \{150, 175, 275\}$, $D_{ij} = \{200, 100, 300\}$
BCTP_4	$c_{ij} = \{6, 4, 1;\ 3, 8, 7; 4, 4, 2\}$ $S_{ij} = \{50, 40, 60\}$, $D_{ij} = \{20, 95, 35\}$
BCTP_5	$c_{ij} = \{10, 2, 20, 11;\ 12, 7, 9, 20; 4, 14, 16, 18\}$ $S_{ij} = \{15, 25, 10\}$, $D_{ij} = \{5, 15, 15, 15\}$
BCTP_6	$c_{ij} = \{9, 12, 9, 6, 9, 10;\ 7, 3, 7, 7, 5, 5; 6, 5, 9, 11, 3, 1; 6, 8, 11, 2, 2, 10\}$ $S_{ij} = \{5, 6, 2, 9\}$, $D_{ij} = \{4, 4, 6, 2, 4, 2\}$
BCTP_7	$c_{ij} = \{4, 6, 9, 5;\ 2, 6, 4, 1; 5, 7, 2, 9\}$ $S_{ij} = \{16, 12, 15\}$, $D_{ij} = \{12, 14, 9, 8\}$
BCTP_8	$c_{ij} = \{4, 3, 5;\ 6, 5, 4; 8, 10, 7\}$ $S_{ij} = \{90, 80, 100\}$, $D_{ij} = \{70, 120, 80\}$
BCTP_9	$c_{ij} = \{4, 1, 2, 4, 4;\ 2, 3, 2, 2, 3; 3, 5, 2, 4, 4\}$ $S_{ij} = \{60, 35, 40\}$, $D_{ij} = \{22, 45, 20, 18, 30\}$
BCTP_10	$c_{ij} = \{5, 7, 10, 5, 3;\ 8, 6, 9, 12, 14; 10, 9, 8, 10, 15\}$ $S_{ij} = \{5, 10, 10\}$, $D_{ij} = \{3, 3, 10, 5, 4\}$

Appendix D: Data set of unbalanced classical TP examples

Problem No. [33, 36, 46, 47]	Data
UCTP_1	$c_{ij} = \{10, 8, 4, 3;\ 12, 14, 20, 2; 6, 9, 25, 25\}$ $S_{ij} = \{500, 400, 300\}$, $D_{ij} = \{250, 350, 600, 150\}$
UCTP_2	$c_{ij} = \{6, 10, 14;\ 12, 19, 21; 15, 14, 17\}$ $S_{ij} = \{50, 50, 50\}$, $D_{ij} = \{30, 40, 55, 25\}$
UCTP_3	$c_{ij} = \{12, 10, 6, 13;\ 19, 8, 16, 25; 17, 15, 15, 20; 23, 22, 26, 12\}$ $S_{ij} = \{150, 500, 600, 225\}$, $D_{ij} = \{300, 500, 75, 100\}$
UCTP_4	$c_{ij} = \{5, 8, 6, 6, 3;\ 4, 7, 7, 6, 5; 8, 4, 6, 6, 4\}$ $S_{ij} = \{800, 500, 900\}$, $D_{ij} = \{400, 400, 500, 400, 800\}$
UCTP_5	$c_{ij} = \{6, 1, 9, 3;\ 11, 5, 2, 8; 10, 12, 4, 7\}$ $S_{ij} = \{70, 55, 70\}$, $D_{ij} = \{85, 35, 50, 45\}$
UCTP_6	$c_{ij} = \{5, 4, 8, 6, 5;\ 4, 5, 4, 3, 2; 3, 6, 5, 8, 4\}$ $S_{ij} = \{600, 400, 1000\}$, $D_{ij} = \{450, 400, 200, 250, 300\}$
UCTP_7	$c_{ij} = \{10, 15, 12, 12;\ 8, 10, 11, 9; 11, 12, 13, 10\}$ $S_{ij} = \{200, 150, 120\}$, $D_{ij} = \{140, 120, 80, 220\}$
UCTP_8	$c_{ij} = \{5, 6, 9;\ 3, 5, 10; 6, 7, 6; 6, 4, 10\}$ $S_{ij} = \{100, 75, 50, 75\}$, $D_{ij} = \{70, 80, 120\}$
UCTP_9	$c_{ij} = \{3, 4, 6;\ 7, 3, 8; 6, 4, 5; 7, 5, 2\}$ $S_{ij} = \{100, 80, 90, 120\}$, $D_{ij} = \{110, 110, 60\}$

UCTP_10	$c_{ij} = \{7, 8, 11, 10;\ 10, 12, 5, 4; 6, 11, 10, 9\}$ $S_{ij} = \{30, 45, 35\}$, $D_{ij} = \{20, 28, 19, 33\}$

Appendix E: Data set of balanced and unbalanced prohibited route TP examples

Problem No. [49, 52]	Data
BPRTP_1	$c_{ij} = \{5, 4, 4, 6, 7, 7;\ 8, 6, 6, 5, 7, 7;\ 7, 6, 6, 8, 9, 9; M, M, 0, 0, M, 0\}$ $S_{ij} = \{100, 200, 150, 110\}$, $D_{ij} = \{80, 40, 50, 60, 80, 250\}$
UPRTP_1	$c_{ij} = \{14, 15, 16, 17;\ M, 16, 17, 18; M, M, 15, 16; M, M, M, 17\}$ $S_{ij} = \{40, 50, 30, 50\}$, $D_{ij} = \{20, 30, 50, 40\}$
UPRTP_2	$c_{ij} = \{10, 13, 16, 19;\ M, 10, 13, 16;$ $M, 15, 18, 21; M, M, 15, 18; M, M, 20, 23; M, M, M, 15\}$ $S_{ij} = \{700, 700, 200, 700, 200, 700\}$, $D_{ij} = \{300, 700, 900, 800\}$
UPRTP_3	$c_{ij} = \{10, 11, 12;\ 15, 16, 17;$ $M, 10, 11; M, 15, 16; M, M, 10; M, M, 15\}$ $S_{ij} = \{300, 100, 300, 100, 300, 100\}$, $D_{ij} = \{200, 400, 300\}$
UPRTP_4	$c_{ij} = \{5.0, 5.5; 6.25, 6.75; M, 5.0; M, 6.25\}$ $S_{ij} = \{920, 250, 920, 250\}$, $D_{ij} = \{800, 1400\}$

Appendix F: Data set of randomly generated TP examples

Problem No.	Data
CTP_1	$c_{ij} = \{19, 16, 17;\ 13, 13, 9; 7, 17, 19\}$ $S_{ij} = \{100, 50, 100\}$, $D_{ij} = \{25, 175, 50\}$
CTP_2	$c_{ij} = \{13, 15, 11, 17;\ 9, 21, 22, 10; 5, 26, 21, 10\}$ $S_{ij} = \{50, 300, 250\}$, $D_{ij} = \{25, 175, 100, 300\}$
CTP_3	$c_{ij} = \{4, 2, 4, 1;\ 3, 10, 5, 1; 4, 8, 8, 6; 4, 10, 2, 7\}$ $S_{ij} = \{400, 300, 250, 50\}$, $D_{ij} = \{100, 300, 250, 350\}$
CTP_4	$c_{ij} = \{37, 30, 19;\ 38, 87, 51; 38, 74, 78\}$ $S_{ij} = \{25, 125, 100\}$, $D_{ij} = \{100, 50, 150\}$
CTP_5	$c_{ij} = \{2, 86, 66;\ 31, 29, 75; 43, 23, 48; 3, 100, 31\}$ $S_{ij} = \{700, 100, 200, 200\}$, $D_{ij} = \{500, 300, 200\}$
PRTP_1	$c_{ij} = \{12, 59, 65;\ 68, 3, 82; M, 88, 80\}$ $S_{ij} = \{100, 150, 50\}$, $D_{ij} = \{150, 65, 85\}$
PRTP_2	$c_{ij} = \{30, 53, 29;\ M, 80, 84; 20, M, 71; 25, 35, 40\}$ $S_{ij} = \{400, 150, 50, 200\}$, $D_{ij} = \{200, 250, 350\}$

PRTP_3	$c_{ij} = \{8, 10, 11, 18;\ M, 7, 9, 17; M, M, 12, 19; M, M, M, 12\}$ $S_{ij} = \{15, 15, 60, 10\}$, $D_{ij} = \{10, 20, 40, 30\}$
PRTP_4	$c_{ij} = \{20, 25, 40;\ 90, 75, 50; M, 30, 35; M, M, 40\}$ $S_{ij} = \{30, 70, 60, 40\}$, $D_{ij} = \{60, 40, 50\}$
PRTP_5	$c_{ij} = \{99, 96, M;\ 52, M, 58; M, 53, 80\}$ $S_{ij} = \{50, 100, 250\}$, $D_{ij} = \{150, 375, 25\}$

Publisher: Eliva Press Global Ltd

Email: info@elivapress.com

Made in United States
North Haven, CT
04 November 2026

21772752R00065